分层空调系统研究与设计思考

Stratified Air Distribution Systems: Research and Design Considerations

程远达　著

中国建筑工业出版社

图书在版编目（CIP）数据

分层空调系统研究与设计思考＝Stratified Air Distribution Systems：Research and Design Considerations：英文/程远达著．—北京：中国建筑工业出版社，2017.1

ISBN 978-7-112-20221-8

Ⅰ.①分… Ⅱ.①程… Ⅲ.①空调设计-英文 Ⅳ.①TB657.2

中国版本图书馆 CIP 数据核字（2017）第 004261 号

分层空调系统研究与设计思考

Stratified Air Distribution Systems：

Research and Design Considerations

程远达　著

*

中国建筑工业出版社出版、发行（北京海淀三里河路 9 号）

各地新华书店、建筑书店经销

唐山龙达图文制作有限公司制版

廊坊市海涛印刷有限公司印刷

*

开本：787×1092 毫米　1/16　印张：8½　字数：211 千字

2017 年 6 月第一版　　2017 年 6 月第一次印刷

定价：**32.00** 元

ISBN 978-7-112-20221-8

（29707）

In this book, the thermal and airflow processes of stratified air distribution (STRAD) systems are experimentally and numerically illustrated. The applications of STRAD systems with separate locations of return and exhaust grilles in three different spaces, namely office space with low ceiling level (lower than 3 m), high space with middle ceiling height (5 m) and large space with very high ceiling level (8 m), were numerically investigated with regard to both thermal comfort and energy saving. Several crucial design parameters for STRAD systems were studied, including the locations of diffusers, the occupied zone cooling load that will be used to determine the required supply air flow rate, and the cooling coil load that will be used to determine the equipment size, and significant results were obtained. Based on the findings in this book, special considerations are also suggested to consultant engineers when designing a STRAD system.

责任编辑：辛海丽
责任设计：李志立
责任校对：陈晶晶　张　颖

Foreword

Stratified air distribution (STRAD) systems have better ventilation effectiveness and are more energy-saving, in comparison with the conventional mixing ventilation systems. The primary aim of this book is to fill several key knowledge gaps for the design of STRAD systems, including the thermal comfort evaluation, the occupied zone cooling load calculation for the air flow rate determination, the cooling coil load calculation, and the locations of diffusers.

The book mainly consists of five parts: 1) comparison of two representative thermal comfort models, which are used to evaluate the thermal environments of STRAD systems, by coupling with computational fluid dynamic (CFD) simulations; 2) development of a cooling coil load calculation method and for the first time to conduct an experimental study in a full-scale chamber ventilated by a STRAD system with separate locations of return and exhaust grilles; 3) numerical investigation on the performances of STRAD systems with different heights of return grille location in a small office environment; 4) optimization of thermal comfort and maximization of energy saving for a lecture theatre with terraced floor conditioned by STRAD systems with different diffuser locations; 5) numerical assessments of alternative stratified air distribution designs for a terminal building with large floor area and high ceiling in terms of thermal comfort and energy efficiency.

A database of effective cooling load factors for different heat sources in three common types of space is built based on a total of 56 additional simulation cases. The database can be adopted by engineers to calculate the occupied zone cooling load and then determine the required supply airflow rate properly. The developed cooling coil load calculation method is concluded and special design considerations for STRAD systems are also presented. This book is expected to be helpful to design the stratified air distribution system.

Contents

Nomenclature

Variable	Description	Unit
a	Linear regression constants	ND
A_i	Cross sectional areas of at the upper boundary of the occupied zone	m^2
b	Linear regression constants	ND
C_e	Contaminant concentration in the extract air	ND
C_{in}	Inlet contaminant concentration	ND
C_{local}	Local contaminant concentration by volume	ND
C_{mixed}	Contaminant if perfectly mixed	ND
C_p	Air specific heat capacity	J/(kg • K)
C_r	Average contaminant concentration in the space	ND
C_s	Contaminant concentration in the supply air	ND
E_z	Zone air distribution effectiveness	ND
F_{sa}	Lighting special allowance factor	ND
F_{ul}	Lighting use factor	ND
$H_{r\text{-}plame}$	Enthalpy flow through the horizontal plane at the upper boundary of the occupied zone	W
H_{sup}	Enthalpy flow of the supply air through all the supply diffusers	W
I	Air exchange rate	L/s
i	A type of heat source	ND
$\dot{m}_e$	Exhaust air flow rate	kg/s
$\dot{m}_o$	Outdoor air flow rate	kg/s
$\dot{m}_r$	Return air flow rate	kg/s
Q	Volumetric flow rate	m^3/s
Q_{coil}	Cooling coil load	W
ΔQ_{coil}	Cooling coil load reduction	W
$Q_{comvective,i\text{-}occupied}$	Convection heat to the indoor air transferred directly from heat source i located in the occupied zone	W
$Q_{conv,surface\text{-}occupied}$	Convection heat to the indoor air transferred from the furniture and interior room surfaces located in the occupied zone	W

Variable	Description	Unit
$Q_{\text{i-occupied}}$	Occupied zone cooling load caused by heat source i	W
$Q_{\text{i-space}}$	Space cooling load caused by heat source i	W
Q_{occupied}	Occupied zone cooling load	W
Q'_{occupied}	Effective occupied zone cooling load	W
Q_{space}	Space cooling load	W
$Q_{\text{un-occupied}}$	Unoccupied zone cooling load	W
Q_{vent}	Ventilation load	W
R_{a}	Outdoor airflow rate required per unit area	m^3/s
R_{p}	Outdoor airflow rate required per person	m^3/s
R_{T}	Total insulation	$(m^2 \cdot K)/W$
T_{f}	Minimum temperatures at floor level	℃
T_i	Inlet air temperature	℃
T_{o}	Outlet air temperature	℃
t	Air temperature	℃
$t_{1.1\text{m}}$	Mass-weight Air temperature at 1. 1m height	℃
t_{e}	Exhaust air temperature	℃
t_{eq}	Equivalent temperature	℃
$\Delta t_{\text{head-ankle}}$	Temperature difference between head and ankle levels	℃
t_i	Temperatures of the nodal points at the upper boundary of the occupied zone	℃
t_{o}	Outdoor air temperature	℃
t_{r}	Return air temperature	℃
t_{s}	Supply air temperature	℃
t_{set}	Room set-point temperature	℃
t_{skin}	Skin surface temperature	℃
V	Volume of space	m^3
V_{bz}	Outdoor airflow required in the breathing zone	m^3/s
V_{oz}	Zone outdoor airflow	m^3/s
V_{s}	Supply airflow rate	m^3/s
w_i	Vertical velocities	m/s
W	Total light wattage	W
Greek symbols		
ε	Contaminant removal effectiveness	ND
ε_{t}	Ventilation effectiveness for heat distribution	ND
ε_{I}	Air change effectiveness	ND
τ	Time constants	s
θ_{age}	Length of time	s

List of Abbreviations

Variable	Description
ACH	Air changes per hour
AET	Average equivalent temperature
AHU	Air-handing unit
CBE	Center for the build environment
CC	Cooled ceiling
CFD	Computational fluid dynamic
COP	Coefficients of performance
DV	Displacement ventilation
Do	Discrete ordinates
DR	Draft rate
ECLF	Overall effective cooling factor
$ECLF_i$	Effective cooling load factor of heat source *i*
EHGFs	Effective heat gain factors
EHT	Equivalent homogeneous temperature
ESHG	Effective sensible heat gain
HVAC	Heating, ventilation and air conditioning
IAQ	Indoor air quality
LTS	Local thermal sensation
MTV	Mean thermal vote
MV	Mixing ventilation
OTTV	Overall thermal transfer value
PAU	Primary air-handing unit
PD	Percentage of people predicted to be dissatisfied due to local discomfort
PPD	Predicted percentage of dissatisfied
PV	Personalized ventilation
RMP	Round movable panel
SBS	Sick building syndrome
SET	Standard effective temperature
SH	Stratification height
SPF	Supply plenum fraction
RCLR	Room cooling load ratio

RPF	Return plenum fraction
RTS	Radiant time series
TSV	Thermal sensation vote
UCB model	UC Berkeley thermal comfort model
UCLR	Cooling load ration
UFAD	Under-floor air distribution
VE	Ventilation effectiveness
ZF	Zone fraction

Chapter 1 Introduction

1.1 Background

The main mission of heating, ventilation, and air conditioning (HVAC) system is to provide a comfortable thermal environment with healthy indoor air quality (IAQ) for the occupants. However, since the variable-air volume systems were first introduced 40 years ago, the HVAC approaches have changed little to keep pace with the continuously improved requirements of living or working environmental quality of occupants. There was a large portion of occupants complained about the thermal comfort and indoor air quality of existing building environments [Huizenga et al. 2006, Bonnefoy et al. 2004], which was disadvantageous to the productivity and health of occupants [Fisk 2000]. In addition, the energy crisis occurred in the 1970s urged the energy saving of buildings to be much more important in recent decades. According to the Hong Kong Energy End-use Data [HKEEUD 2011], the HVAC system has been becoming one of the main energy consumers in buildings. Therefore, a broader effort is called for to promote the energy efficiency of the HVAC system.

Stratified air distribution (STRAD) systems, including displacement ventilation (DV) systems and under-floor air distribution (UFAD) systems, have attracted more and more attentions over the past ten years due to its better ventilation efficiency, in comparison with the conventional well-mixed ventilation systems. In a STRAD system, the cold, fresh air is supplied directly to the occupied zone and flows up due to thermal buoyancy. The free convection flow around human body may protect the breathing zone from surrounding contamination at the head level. Thus, the inhaled air quality is improved [Brohus and Nielsen 1996, Bauman 2003]. The zone air distribution effectiveness in STRAD systems has been demonstrated to be higher than that in mixing ventilation (MV) systems [ASHRAE 2011]. Previous studies also claimed that STRAD systems were able to provide superior thermal comfort compared with conventional MV systems [Bauman 2003, Seppanen et al. 1989]. In STRAD systems the thermal neutral temperature has been demonstrated to be 2.5℃ higher than that in MVsystems based on an experimental study, and satisfied thermal environment was maintained up to 27℃ for the room air temperature in Hong Kong [Fong et al. 2011]. Furthermore, in STRAD systems only the lower, occupied zone of a space need to be cooled while leaving the upper zone uncooled [Bagheri and Gorton 1987]. Thus, the STRAD system inherently provides potentials to save energy.

For the advantages of STRAD systems, they have been achieved considerable acceptance in Europe, South Africa, Japan and North America. However, the air flow in a STRAD system is much more complicated than that in a MV system. It demands more rigorous analysis of building thermal and airflow processes, which may be beyond the capability of most practicing consulting engineers. On the other hand, as a relatively new technology, there is still lack of information on several key design issues of STRAD systems, for instance the reliable occupied zone cooling load calculation methods and corresponding equipment size determination methods, which hinder the wide applications of STRAD systems. As described by Bauman [2003] "UFAD technology is now in a situation where systems are being designed and installed at an increasingly rapid pace, even before a full understanding and characterization of some of the most fundamental aspects of UFAD systems performance have taken place." Furthermore, additional energy saving potential is able to be provided for STRAD system by splitting the locations of return and exhaust grilles [Lau and Niu 2003, Xu et al, 2009]. However, researches on STRAD systems with similar configurationare still at preliminary stage. Rare information was provided on the evaluation and optimization of the energy saving for STRAD systems with separate locations of return and exhaust grilles. Therefore, for the purpose of developing some fundamental understandings on the design of STRAD systems, especially for system with separate locations of return and exhaust grilles, further in-depth studies are required.

1.2 Objectives

The challenge for a team to design a STRAD system is to capture as many of the benefits as possible while keeping the negative influences to a minimum. Therefore, this book aims to clearly illustrate the potential advantages of STRAD systems with separate locations of exhaust and return grilles, and conceive to fill the knowledge gaps that confusing the consultant engineers.

The following objectives are going to be achieved in this book:

• To assessthe reliability and stability of two representative thermal comfort models, which are able to evaluate thermal comfort in non-uniformenvironments, by coupling them with computational fluid dynamic (CFD) simulations.

• To develop a cooling coil load calculation method and validate it in experimental and numerical studies.

• To demonstrate that splitting the locations of return and exhaust grilles is able to significantly improve the energy efficiency of STRAD systems through an experimental study.

• To illustrate the energy saving principles in STRAD systems with separate locations of return and exhaust grilles, when applied in three common types of spaces, through numerical studies.

- To clarify the influence of diffusers locations on the performance of STRAD systems with separate locations of return and exhaust grilles.
- To build up a database that can be adopted by engineers to calculate the occupied zone cooling load and then determine the required supply airflow rate.

1. 3 Structure of the book

This book consists of eight chapters and the contents of each chapter are summarized as follows:

Chapter 1 states the background and objectives of this study, and provides an overview of the book.

Chapter 2 reviews the advantages of STRAD systems first. It is followed by a description of the evaluation criteria for the performance of STRAD systems, in terms of indoor air quality, thermal comfort, and energy consumption. The following part introduces several key design parameters that strongly affect the performance of STRAD systems. A closure of this chapter is made by the importance of this book and a summary section.

Chapter 3 couples two representative thermal comfort models with CFD simulations to evaluate stratified thermal environments. The evaluation results are compared in terms of reliability and stability. Innovative coupling procedure between the thermal comfort model and the CFD simulationis proposed with simplified operations and improved accuracy.

Chapter 4 clarifies the energy saving principle for STRAD systems and illustrates the extra energy saving potential provided by splitting the locations of return and exhaust grilles. A novel CFD-based cooling coil load calculation method is developed, which can be used to evaluate the energy saving in STRAD systems and determine the equipment size. For the first time, experimental study in a full-scale chamber ventilated by STRAD systems with separate locations of return and exhaust grilles is conducted to validate the cooling coil calculation method.

Chapter 5 conducts a CFD validation process first. Then, the application of STRAD systems with separate locations of return and exhaust grilles in a small office hypothetically located at perimeter zone is numerically investigated. The influence of return grille location on the performance of STRAD systems is assessed in terms of thermal comfort and energy saving. Effective cooling load factors for each heat sources in the space are calculated.

Chapter 6 numerically studies different designs to realize thermal stratified air distribution in a large-space lecture theatre with terraced-floor. The thermal environment is optimized and the energy efficiency of the STRAD system is maximized. The impact of return grilles locations on the thermal environment and energy efficiency of STRAD systems is also investigated. Effective cooling load factors for each heat sources are provided to calculate the occupied zone cooling load in STRAD systems, when applied in a large space lecture theatre with similar configuration.

Chapter 7 investigates two typical practical air distribution designs in a terminal building with large floor area and high ceiling. The thermal environments and energy saving potentials for these two ventilation designs are compared. The impacts of the supply and return diffusers locations on the ventilation performance are studied. The influences of solar radiation on the thermal distribution and energy efficiency of different ventilation systems are also illustrated.

Chapter 8 summarizes the conclusions of the present work and provides some recommendations for future work. Special considerations on the design of STRAD system are also provided.

Chapter 2 Literature Review

2.1 Introduction

After various attempts in the early 19th century, air conditioning was invented by Willis Haviland Carrier in the year of 1902. Carrier had his first patent "Apparatus for Treating Air" (U. S. Patent808897) granted in 1906, but it was a textile engineer Stuart H. Cramer who first used the term 'air conditioning' in a 1906 patent claim [McQuiston and Parker 1994]. Since then, the heating, ventilation, and air conditioning (HVAC) system has become one of the most important portions in buildings and developed rapidly during the century. However, the never-ending complaints from occupants about thermal comfort and indoor air quality (IAQ) are still the main issues for the ventilation of modern buildings, which result in reduced productivity and huge economic loss. It was revealed that an average productivity loss of 3% was attributed to poor IAQ, based extensive surveys of 94 state government office buildings in the New England area [Axelrad 1989]. The annual economic impact related to the respiratory disease, allergies and asthma, and sick building syndrome (SBS) was estimated to be $20 to $200 billion for the United States [Fisk 2000]. Thus, it is not sufficient for the HVAC system to supply air to the building spaces with the correct temperature and flow rate, but it is also called for to design the ventilation systems in such a way that occupants experience high air quality and thermal comfort.

In ASHRAE Handbook [ASHRAE, 2011], the room air distribution systems are categorized as follows: Conventional mixed systems (e. g. , overhead distribution)——which have little or no thermal stratification of air within the occupied and/or process space; Full thermal stratification systems (e. g. , thermal displacementventilation)——which have little or no air mixing in the occupied and/or process space; Partially mixed systems (e. g. , most under-floor air distributiondesigns)——which provide limited air mixing in the occupied and/or process space; Task/ambient air distribution (e. g. , personally controlled deskoutlets, spot conditioning systems)——which focus on conditioning only part of the space for thermal comfort and/or process control. Figure 2. 1 presentsthe air flow characteristics within different air distribution systems. In a conventional mixed system, as shown in Figure 2. 1 (*a*) the air is supplied at ceiling level with high-velocity and induces surrounding air sharply. For a well-designed mixed ventilation system, the air temperature and contaminant distribution in the space is uniform.

Different with that, in typical thermal displacement ventilation (DV) system, the air is supplied from low sidewall diffusers with a slightly lower temperature than the desired room air temperature and entertained very little room air. Because the supply air is colder than room air, it is spread over the floor and heated by the heat sources in the occupied zone. As presented in Figure 2.1 (*b*), free convection from heat sources creates a vertical air movement in the room, and the heated air is removed by return openings located in the ceiling or just below the ceiling. Vertical temperature gradient is the main characteristic of a room ventilated by a DV system. The air flow characteristic in the underfloor air distribution (UFAD) system is falling between a fully mixed system and a fully stratified system. In an UFAD system, the supply air is discharged, usually vertically, at relatively high velocities and entrains room air in a similar fashion to outlets used in mixed air systems. Thus, the general flow in the room is controlled not only by the free convention or the buoyancy but also by the momentum flux from the diffusers. It is the primary difference between UFAD to DV that the general flow is governed by the free convention or the buoyancy only [McDonell 2003].

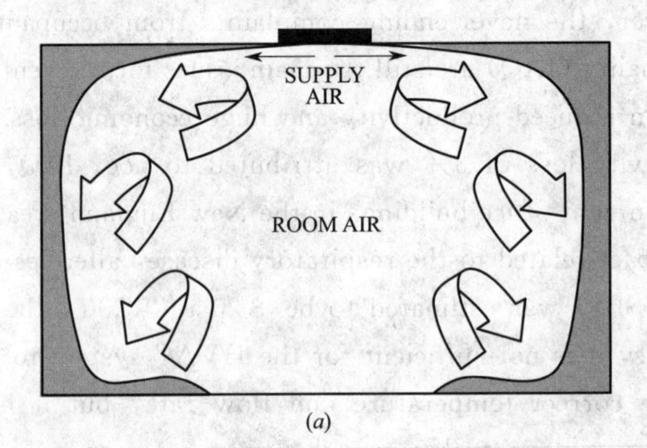

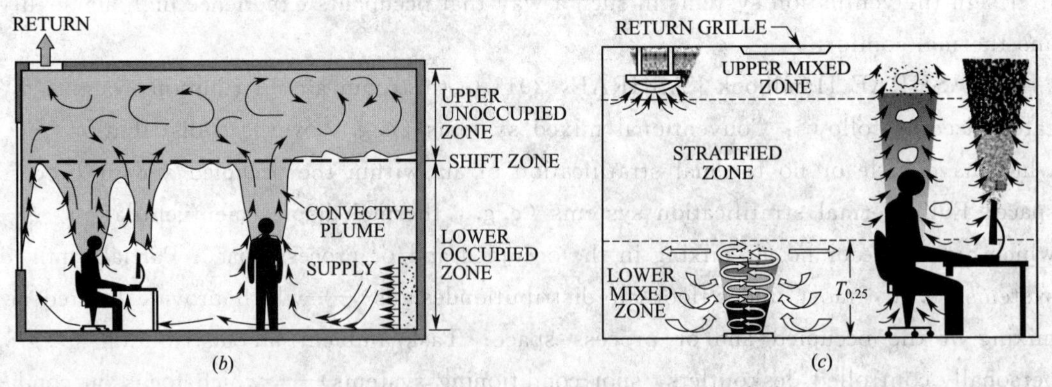

Figure 2.1 Different air distribution system

(*a*) Conventional mixed system; (*b*) Thermal Displacement ventilation; (*c*) Under Floor air distribution

As shown in Figure 2.1 (*c*), there are three distinct zones formed in a typical UFAD system: the lower mixed zone, the middle stratified zone and the upper mixed zone [ASHRAE 2011]. Stratified air distribution (STRAD) systems have received more and more attentions due to the claims of improved indoor air quality and reduced energy consumption. A STRAD system was defined as a means of providing supply air directly to the occupants in a space and cooling the lower, occupied zone of a building while leaving the upper zone uncooled [Gorton and Bagheri 1986]. Hence, the STRAD system may include any kind of system which maintains a stratified temperature and contaminant distribution in a space. Displacement ventilation (DV) and Under Floor air distribution (UFAD) are two typical STRAD systems used broadly in practical. During the past decades, the STRAD systems have been applied world widely and successfully not only in industrial buildings but also in commercial buildings. It was estimated in 1989 that displacement ventilation accounted for a 50% market share in industrial applications and 25% in office applications in Nordic countries [Svensson 1989]. While in the U.S, about 12% of new offices used raised floors by 2005 and half of which incorporated with UFAD technology [Liu 2005]. In Hong Kong, several new buildings, including The Cheung Kong Center, The HSBC headquarter and The Center in Central, are also ventilated by the STRAD systems. With the improvement in the technology and the accumulation on the design experience, it is believed that the STRAD systems are going to be achieved much wideracceptance in new buildings.

2.2 Performances of stratified air distribution systems

2.2.1 Indoor air quality

The main motivation to develop STRAD systems is to reduce the risk of SBS and achieve better indoor air quality for occupants. Thus, the ventilation effectiveness of STRAD systems has been widely studied since they were proposed. In a STRAD system, the cold fresh air is supplied directly to the occupied zone with a slightly lower temperature than that of the room air. The air flows up due to thermal buoyancy when it heated by heat sources. Therefore the contaminants, which related to heat sources and located at the upper part of the room, is possible to be extracted directly from the exhaust inlets. Furthermore, the free convection flow around a person may induce the fresh air from the lower part of room to the breathing zone of occupants and protect the breathing zone from surrounding contamination. This advantage has been already confirmed in previous experimental studies [Brohus and Nielsen 1996]. High ventilation effectiveness has been testified to be available for a properly designed STRAD system [Chen and Glicksman 2003, Bauman 2003]. As shown in Figure 2.2, ASHRAE Standard 2011 presents the CO_2 concentration distributions in a roomventilated by four different air distribution systems, which varied from fully stratified air distribution systems to fully mixed room air distribution systems. It is obviously that compared with the MV systems, the STRAD systems provide better indoor air quality for occupants. In the table 6-2 of ASHRAE Standard 62.1

[2010], the zone air distribution effectiveness E_z is presented, but with fixed value for several types of air distribution systems. The value is 1.0 for most UFAD systems and 1.2 for DV systems. Because the zone air distribution effectiveness plays an important role in determining the minimum amount required of outdoor air for a space to achieve satisfied IAQ, it is considerable to use an accurate E_z value for stratified air distribution systems. For this purpose, a database that contained the simulation results of 102 cases was established by Lee et al. [2009a]. Based on that, a set of correlation equations that can be used to calculate the zone air distribution effectiveness in different conditions was also developed.

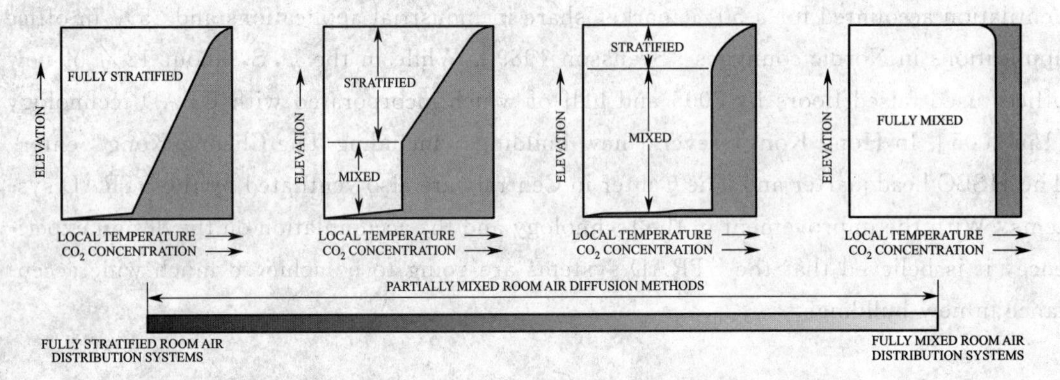

Figure 2.2 The temperature and concentration distribution within different air distribution systems [ASHRAE 2011]

However, previous studies revealed that DV systems might not always provide better IAQ than MV if the contaminant sources were not associated with heat sources, for example VOCs emitted from building materials [Lin et al. 2005a, Cheong et al. 2006], or the thermal plumes generated by heat source was too weak to bring contaminants to the ceiling [Stymne et al. 1991, Mundt 1996]. Nielsen [1993] pointed out as well that the down draft caused by a cold wall or window was possible to bring the polluted air from the upper zone to the lower zone and reduce the ventilation efficiency of STRAD systems. The thermal plumes from occupants were also possible to attract contaminant from a passive source with a vertical location in the room below the breathing zone of the person, and reduce the ventilation efficiency [Yu et al. 2009]. Moreover, the moving of occupants also strongly affected the distribution of contaminants in STRAD systems [Mattsson and Sandberg 1994]. Recently based on a numerical study, Tian et al. [2010] concluded that the ventilation effectiveness of the STRAD system significantly depended on the positions of pollutant sources and heat sources. Therefore, when designing a stratified air conditioning system, it needs to be considered comprehensively for the layouts of the indoor heat and pollution sources.

2.2.2 Thermal comfort

The thermal environment greatly influences the productivity of occupants, therefore it is always the research emphasis of ventilation systems. Gan [1995] has conducted a simulation research on the local thermal discomfort in an office room. The Fanger's comfort equations were adopted to evaluate the thermal comfort level and draft risk. The results indicated that thermal discomfort could be avoided by optimizing the supply conditions. Nielsen et al. [2003] made a comparison study on the thermal comfort performance of DV and MV based on a maximum velocity assumption and a restricted vertical temperature gradient in the room. He supposed that with controlled maximum velocity and temperature gradient, the same comfort level could be obtained independent of the air distribution system. Similar with that, Lin et al. [2005b] found that when properly designed, a thermally comfortable environment with a low air velocity, a small temperature difference between the head and ankle levels, and a low percentage of people dissatisfied could be maintained in both DV and MV systems. Moreover, several studies reported that thermal comfort of occupants may benefit from using the manually adjustable diffusers in UFAD systems, since the personal environmental control is available. A previous field study [Bauman et al. 1995] concluded that the occupants who were given the ability to control local environment were almost twice as tolerant of temperature differences and thus expressed fewer thermal complaints. Lehrer and Bauman [2003] pointed out that "When occupants are given control of their environment, they are much more tolerant of a range of conditions" . Subjective survey results collected by the Center for the Build Environment (CBE) also revealed that UFAD could lead to a higher degree of satisfaction, as 88% of panelists expressed preference for the thermal environment with the studied UFAD system over that with an overhead MV and 50% of them reported floor air supply enhanced their ability to perform their work [Webster et al. 2002]. However, due to non-uniform thermal environment in stratified air distribution (STRAD) systems, thermal comfort problems are still required to be considered carefully in the design stage. The draught discomfort and temperature stratification problems associated with floor-based ventilation systems have been revealed from the experimental results conducted by Arens et al. [1991]. These were two main issues that caused discomfort in a STARD system [Cho et al. 2005, Melikov and Nielsen 1989], which should be avoided in the design.

2. 2. 3 Energy consumption

According to the Hong Kong energy end-use data [HKEEU data 2011], 14% of all the energy consumed in Hong Kong at 2009 was used by HVAC systems in the building sector. The statistics revealed that a broader effort is called for to promote the energy efficiency of the HVAC systems. Thus, comprehensive studies have been conducted on the energy performance of STRAD systems. As shown in Figure 2. 3, Bauman [2003] presented typical vertical temperature profiles in a room ventilated by UFAD systems, DV systems, and conventional MV systems, respectively. It indicates that the thermal stratification is obvious both in UFAD and DV systems, which is benefit to the energy saving of the system, since convective heat sources at or above the stratification height will rise up and exit the space without mixing into the lower.

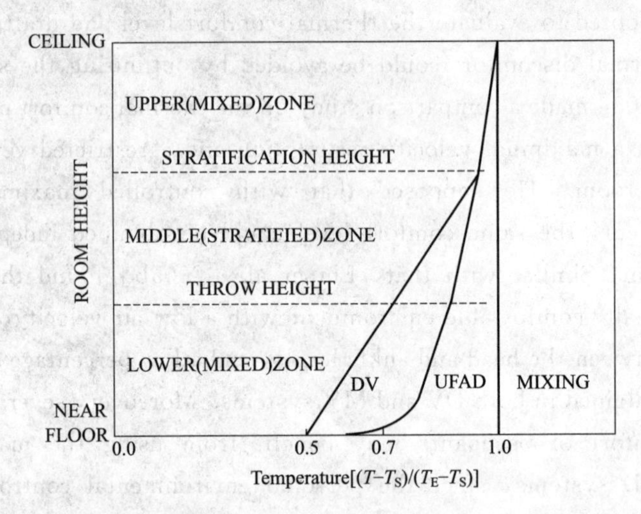

Figure 2.3 Comparison of typical vertical temperature profiles for under-floor air distribution, displacement ventilation, and mixing systems [Bauman, 2003]

Based on CFD simulations, Awbi [1998] found that when cooling a room, the ventilation effectiveness for heat distribution in displacement ventilation (DV) system was almost twice of the value in mixing ventilation (MV) system, which meant that displacement ventilation was more energy efficient. The ventilation effectiveness for heat distribution ε_t is similar to heat exchanger effectiveness and defined as:

$$\varepsilon_t = \frac{T_o - T_i}{\overline{T} - T_i} \tag{2-1}$$

where, T is air temperature (℃), $\overline{T}$ is the mean value for the occupied zone. The subscripts i and o refer to inlet and outlet respectively. Hu et al. [1999] compared the energy consumption of a DV system with that of a MV system in different types of buildings. For the DV system, they stated that "a higher supply temperature means that the cold-water temperature can also be higher than that for mixing ventilation. Therefore, the chiller in displacement ventilation can have better performance." However, they found this benefit was climate dependent and the total energy consumption in DV system was slightly less than that in MV system. Different with that, a number of studies [Heinemeier et al. 1990, Int-Hout 2001] identified that significant energy benefits would be obtained in UFAD systems due to high cooling air temperatures, which extended the economizer or free-cooling capabilities of HVAC systems and were especially significant in mild climates. Figure 2.4 presents the range of energy savings that have been found through field studies, simulations, and laboratory studies for UFAD systems. The figure revealed that for different climate types, the energy savings for UFAD systems were varied largely and between 5%~40%. Alajmi and El-Almer [2010] investigated the effectiveness of UFAD systems, when applied in various types of commercial buildings in hot climate. The simulation results revealed that compared with ceiling-based air distribution, a significant energy saving, up to 30%, was available for UFAD system, especially for buildings with high ceiling.

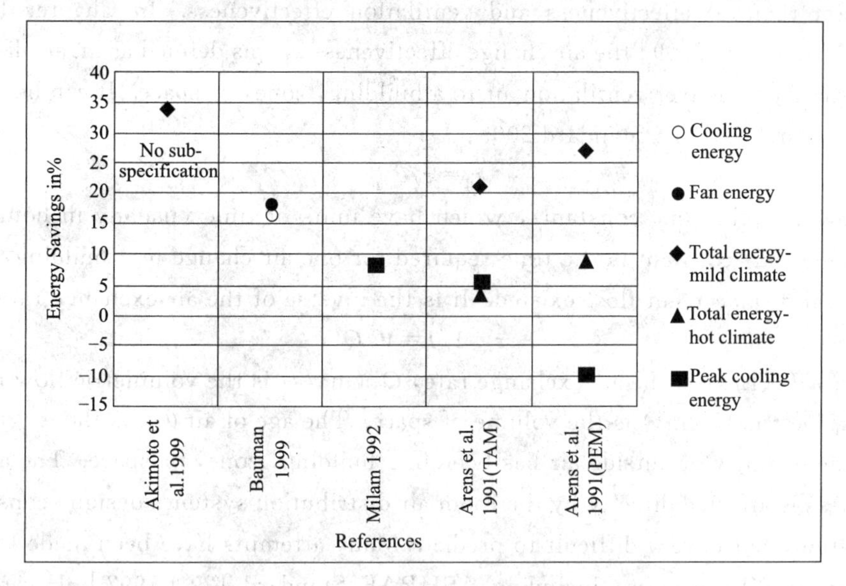

Figure 2. 4 The annual energy savings for underfloor air systems in different researches [Loftness et al. 2002]

Xu and Niu [2006] coupled the CFD simulation and dynamic cooling load simulation to predict the annual energy consumption of an UFAD system. They concluded that the energy saving for an UFAD system derived from three factors: an extended free cooling time, a reduced ventilation load and increased coefficients of performance (COP) for chillers. This conclusion was consistent with that presented by Bauman [2003]. He concluded that the energy savings for UFAD systems over conventional overhead systems were predominately associated with two major factors: (1) cooling energy savings from economizer operation and increased chiller COP and (2) fan energy savings due to reduced total air volume and reduced static pressure requirements. By using TRNSYS, the energy saving capacity of the stratum ventilation, in which the air is supplied at an intermediate level close to the head level, has been investigatedin three typical configurations in Hong Kong [Lin et al. 2011a, Lin et al. 2011b]. The results indicated that the energy saving potential for the stratum ventilation system was prominent and derived from five factors: the reduction in ventilation, dehumidification and transmission loads, prolonged free cooling period and increased the COP of the chillers.

2. 3 Evaluation criteria for stratified air distribution systems

2. 3. 1 Indoor air quality evaluation

As reviewed in section 2. 2. 1, previous studies investigated the ventilation performance of STRAD by using various indices. Generally speaking, there are four most widely used indices of indoor air quality: number of air changes, air change effectiveness, contaminant removal effectiveness and ventilation effectiveness. In Chapter 16 of the ASHRAE Standard [2009] the air change effectiveness ε_I was defined as an air distribution system's ability to deliver ventilation air to a building, zone, or space. It can be calculated as presented in ASHRAE Standard 2009:

$$\varepsilon_I = \tau / \theta_{age} \tag{2-2}$$

where, τ is the time constants, which have units of time (usually in hours or seconds). One time constant is the time required for one air change in a building, zone, or space if ideal displacement flow existed. It is the inverse of the air exchange rate:

$$\tau = 1/I = V/Q \tag{2-3}$$

where, I (L/s) is the air exchange rate, Q (m^3/s) is the volumetric flow rate of air into the space and V (m^3) is the volume of space. The age of air θ_{age} is the length of time t that some quantity of outside air has been in a building, zone, or space. The air change effectiveness is affected directly by the room air distribution system's design, construction, and operation, but is very difficult to predict. Many attempts have been made to quantify the air change effectiveness, including ASHRAE Standard 129 [1997]. However, this standard is only for experimental test conditions in well-controlled laboratories, and should not be applied directly to real buildings.

The contaminant removal effectiveness ε is one of the first indices that actually define a perceived air quality. It is described as a measure of how effectively contaminants that generated in the space are removed from the space and prevented from spreading in the occupied zone, calculated as following equation:

$$\varepsilon = \frac{C_e - C_s}{C_r - C_s} \tag{2-4}$$

where, C_e is contaminant concentration in the extract air, C_r is the average contaminant concentration in the space and C_s is the contaminant concentration in the supply air. The contaminant removal effectiveness was influenced by the supply flow rate, by the thermal loads and by the type of contaminant source [Olesen et al. 1994]. The differential and correlations between air change effectiveness ε_I and contaminant removal effectiveness ε have been numerically studied for typical indoor spaces and ventilation systems. The results indicated that the correlations were mostly depended on ventilation strategy, which were strong for mixing ventilation systems and weaker for displacement ventilation systems [Novoselac and Srebric 2003].

Ventilation effectiveness is a description of an air distribution system's ability to remove internally generated pollutants from a building, zone, or space. ASHRAE Standard [2011] cited the ventilation effectiveness (VE) as a measure of mixing within the volume relative to a perfectly mixed system and is described with the following equation:

$$VE = \frac{C_{mixed} - C_{in}}{C_{local} - C_{in}} \tag{2-5}$$

where, C_{local} is the local contaminant concentration by volume; C_{in} is inlet contaminant concentration; C_{mixed} is contaminant if perfectly mixed.

Therefore, ventilation effectiveness indicates the degree of contaminant stratification with the volume. $VE>1$ means that concentrations in the breathing zone are lower than in a perfectly mixed system; $VE<1$ means they are higher.

In ASHRAE Standard 62.1 [2010], it described "The ability of the ventilation system to deliver outdoor air to the breathing zone of the space can be described by two factors: zone air distribution effectiveness and system ventilation efficiency as applied to multiple space recirculating systems." The Zone Air Distribution Effectiveness E_z was defined as a measure of how effectively the zone air distribution uses its supply air to maintain acceptable air quality in the breathing zone. While the System Ventilation Efficiency was defined as the efficiency with which the system distributes air from the outdoor air intake to the breathing zone in the ventilation-critical zone, which required the largest fraction of outdoor air in the primary air stream. The values of Zone Air Distribution Effectiveness E_z are presented in Table 6-3 of ASHRAE Standard 62.1 [2010]. It was illustrated in the Chapter 6 of ASHRAE Standard 62.1 [2010] that these three indices, i.e., the Zone Air Distribution Effectiveness E_z, Ventilation Effectiveness VE and Air Change Effectiveness ε_I, had slightly different definitions but essentially measure the same effect: the ability of the system to deliver air from the supply air outlet to the breathing zone of the space.

2.3.2 Thermal comfort evaluation

In a STRAD system, individuals exposed to non-uniform environment and significant temperature variations over their bodies. The distributions of local skin temperature, and local sensible heat gain and loss for human body segments in such environments are varied largely. Thus, the thermal comfort evaluation in a STRAD system with asymmetrical environment is more complicated than that in a uniform environment, since thermal sensation is highly dependent on local heat transfer characteristics. Yu et al. [2006] reported a subjective study in a field environmental chamber (FEC) served by a displacement ventilation system. The results indicated that local thermal comfort of body segments were affected by both overall and local thermal sensations. In a recent study, Sun et al. [2012] also found that the whole body thermal sensation of occupants was correlated with the local thermal sensation (LTS) at the waist, the arms, the calf and the feet when a displacement ventilation system was employed. Thus, previous studies [Jones 2002, Zhang et al. 2005] pointed out that the most frequently cited thermal comfort standards: ASHRAE Standard 55 [2010] and ISO 7730 [1994] may not suitable to evaluate thermal comfort in STRAD systems with asymmetrical radiation and local airflow. Even though several limits have been adopted for asymmetrical environments in these standards, for instance, the vertical temperature difference between the ankle and the head levels and the local draft that affected by local turbulence, the overall effects on thermal comfort combined with asymmetrical radiation and local airflow cannot be properly evaluated.

Therefore, a number of updated thermal comfort models that concerning on the asymmetrical environments have been proposed and developed in the past decades [Guan et al. 2003]. The thermal interaction of human with the environment involves both a physiological response and psychological response. Thus, the thermal comfort models can further categorized as physiological or psychological models [Walgama et al. 2006]. Table 2.1 presents several representative human thermal physiological models, which from simplest one-node model [Givoni and Goldman 1971] to the complicated three-dimensional finite element model [Smith 1991, Fu 1995].

Human thermal physiological models **Table 2.1**

Data	Author	Description
1971	Givoni and Goldman	One-node model; Empirical
1971	Gagge et. al	Two-node model
1977	Azer and Hsu	Two-node model
1992	Jones and Ogawa	Two-node model with transient response
1971	Stolwijk	Multi-node thermal model
1985	Wissler	225-node finite element thermal model
1991	Ring and de Dear	40-layer finite difference skin model
1991	Smith	Three-dimensional, finite element model
1995	Fu	
1999; 2001	Fiala D	Multi-node thermal model; 366 tissue nodes
2001	Huizenga C et al.	Multi-node thermal model; arbitrary number of segments
2002	Tanabe et. al	Multi-node thermal model; 65-node

Most multi-segment multi-node physiological models adopted currently were modified and improved from parameters and thermal control equations of Stolwijk model [Stolwijk 1971]. The physiological modules in the IESD-Fiala model [Fiala et al. 2001, Fiala et al. 2010] and the UC Berkeley Comfort Model [Huizenga et al. 2001, Zhang et al. 2001] are two typical representatives of them. These two models were capable to predict physiological responses of different body segments to transient or non-uniform thermal environments, with considerations of the individual physiological differences, such as the body density, metabolic heat production, blood flow rate.

Extensive investigations also have been conducted on physiological models. However, fewer researches considered the thermal sensation and human subjective response to the asymmetrical environment [Fiala 1998]. Table 2.2 presents several thermal sensation and comfort models that devoted to non-uniform environments. Because extreme asymmetries and transients conditions were often experienced in vehicle interiors, thermal sensation evaluation was first widely studied in the automobile industry [Taniguchi et al. 1992, Hagino and Junichiro 1992, Matsunaga et al. 1993]. More recently, a comprehensive thermal sensation and comfort model——The UC Berkeley thermal comfort model (UCB Model), has been developed for a wide range of environments [Zhang 2003, Arens et al. 2006]: uniform and non-uniform, transient and stable. The thermal sensation scale used in the model was extended from ASHRAE-7 point scale, adding "very hot" and "very cold" to accommodate extreme environments. This model included all the major effects that have been observed about human thermal comfort responses to thermal environments, and it was the first model that addressed human responses to simultaneous asymmetrical and transient thermal conditions.

Human thermal physiological models **Table 2.2**

Data	Author	Description
1989	Wyon et al.	Concept of the "Piste"; Steady-state; Non-uniform
1992	Taniguchi et al.	Whole body thermal sensation
1992	Hagino and Junichiro	Statistical model; Non-uniform
1993	Matsunaga et al.	Adopt AET* value to calculate PMV; Overall comfort
1993	de Dear et al.	Dynamic Thermal Stimulus Model; Whole body thermal state
1994	Wang X	Whole body comfort; Transient thermal sensation
2002; 2003	Kohri and Moschida	Local SET*; Uniform; Non-uniform
2003a; 2003b	Guan. et al.	Transient thermal sensation prediction model
2003	Fiala	Dynamic Thermal Sensation model; Transient thermal sensation model
2010	Fiala et al.	
2003	Nilsson and Holmér	New thermal comfort zones based on t_{eq}; Steady-state; non-uniform
2007	Nilsson	
2006	Arens et al.	UC Berkeley Thermal Comfort Model (UCB Model); Transient, Non-uniform or uniform
2010a; 2010b; 2010c	Zhang H et al.	

EHT*: Equivalent Homogeneous Temperature; AET*: Average Equivalent Temperature;
SET*: Standard Effective Temperature; TSV*: Thermal sensation vote;
t_{eq}*: Equivalent temperature

Different with that, Nilsson and Holme'r [2003] adopted the value of equivalent temperature (t_{eq}) as an index to predict thermal comfort in asymmetrical environments, which was similar to the 'Piste' model [Wyon et al. 1989] but with several innovations. As shown in Figure 2.5, the basic assumption of this method is that the human body segments are equally sensitive to the same heat losses independent of the insulation of clothing worn and environment being. The relationship between t_{eq} and the mean thermal vote (MTV) scale was built up by regression of the experimental data together [Nilsson 2007] and prescribed as following:

$$t_{eq}=t_{skin}-R_T(a+bMTV) \tag{2-6}$$

where, t_{skin} is the skin surface temperature (℃); R_T is the total insulation, including resistance of clothing and air layer; a, b are the linear regression constants. Then the clothing independent thermal comfort diagrams were described by calculating the limited t_{eq} values for the corresponding MTV values of each scale zone borders [Nilsson 2007]. The basic assumption of this empirical method may not be true, especially in extreme cases such as no clothing or heavy clothing. It is only suitable for steady conditions, and humidity effects and latent heat effects are not included at all. However, as a convenient way to evaluate thermal comfort in non-uniform environment, the method has been widely quoted and promoted to the ISO Standard 14505 [2007].

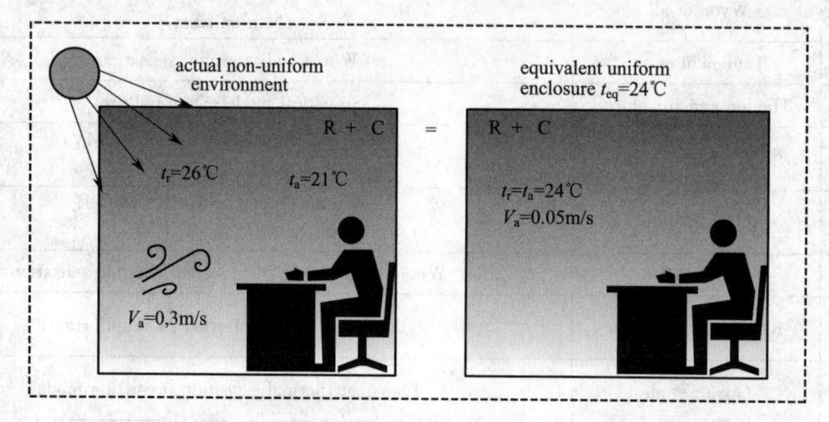

Figure 2.5 The basic assumption of "thermal comfort zone" model [Nilsson and Holme′r 2003]

2.4 Key design parameters for STRAD systems

For a well-designed STRAD system, it can improve indoor air quality, reduce energy use, improve satisfaction and productivity of workers, increase flexibility, and reduce life-cycle costs over conventional mixing ventilation systems [Bauman and Webster 2001, Bauman 2003]. As reviewed in section 2.2, the performance of STRAD systems was strongly affected by several key design parameters, including the space types, diffuser types and positions, the ventilation rate etc. [Lin et al. 2005c, Kobayashi and Chen 2003]. Therefore, a systematic review is called for to be conducted on the impacts from these design parameters to the performance of STRAD systems.

2.4.1 Space types

The STRAD systems can be applied in restaurants, theatres, cinemas, offices, classrooms, meeting rooms etc. However, the space constructions may greatly affect the performance of STRAD systems. It was pointed out that the benefits of STRAD would be more likely realized in buildings with high ceiling and all occupants located at floor level [Gorton and Bagheri 1986]. Previous studies [Skistad 1994, Skistad et al. 2002] also revealed that displacement ventilation was especially effective for buildings with a ceiling height that higher than 3m. They also concluded that the displacement ventilation was usually preferable when contaminants were warmer and/or lighter than the surrounding air. Novoselac and Srebric [2002] reviewed the studies of cooled ceiling and displacement ventilation (CC/DV) systems in buildings. They found that the temperature and contaminant concentration profiles in the occupied zone were almost identical when the ceiling height varied from 2.5m to 5.0m. Lin et al. [2005a, 2005b] compared the performance of a displacement ventilation system with that of mixing ventilation system for different type of spaces, in terms of thermal comfort and indoor air quality. They reported that in the occupied zone, the percentage of people predicted to be dissatisfied due to local discomfort (PD) was generally less than 10% for the office, the classroom and the retail shop with DV, while the value could be greater than 25% for the industrial workshop. They also addressed that the CO_2 concentration for the office, the classroom and the industrial workshop was about 700～800 ppm in most of the space, while the value was higher in the Retail shop. Later, they [Lin et al. 2006] systematically studied the effects of headroom, which defined as the floor to ceiling height of a "floor" in the building, on the performance of DV systems. The research focused on a typical office layout with two different headrooms. The results revealed that the higher the headroom, the lower the mean temperatures in the occupied zone and predicted percentage of dissatisfied (PPD) values for the DV system. Also, a model has been developed to investigate the effect of headroom on the performance of DV in a typical office building. A numerical study [Hashimoto and Yoneda, 2009] was carried out in a modeled room with three different ceiling heights. The results indicated that room with a higher ceiling is beneficial for improving thermal comfort and saving energy for displacement ventilation. Recently, Lee et al. [2009b] also found that in STRAD systems, the spaces with a high ceiling, such as workshops and auditoriums, had a higher value of air distribution effectiveness than those with a low ceiling. The database of air distribution effectiveness was also developed based on numerical simulationsin six different types of indoor space.

2.4.2 Heat source types and locations

As described by Bauman [2003]: "Thermal plumes that develop over heat sources in the room play a major role in driving the overall floor-to-ceiling air motion by entraining air from the surrounding space and drawing it upward." The fluid mechanics of buoyancy air flow driven by a point source of heat was illustrated by Linden et al. [1990]. When rising up due to buoyancy, the thermal plume entrains surrounding air and therefore increases in size and volume, while it gradually decreases in velocity from its maximum just above the heat source. As reviewed by Yuan et al. [1998], the maximum height of the plume was an important design parameter and had significant effects on the ventilation performance. Mundt [1992] has developed an equation to calculate the maximum height of a plume in a space with air temperature gradient and it indicated that the height was only determined by the convective heat emission from heat source. Bauman [2003] supposed that the maximum height of a plume in a space was depended primarily on the heat source strength and secondarily on the stratification in the room. Nielsen [1996] illustrated the effect of different heat source types on the dimensionless temperature at floor level. A design chart as presented in Figure 2.6 was developed, based on experimental results in rooms with heights of 2.5m to 4.5m. The effects of more realistic heat source geometries to the performance of STRAD systems, for example, multiple point sources of heat [Cooper and Linden 1996], fully distributed heat sources [Gladstone and Woods 2001] and human body [Bjorn and Nielsen 1995] also have been investigated with detail. A conclusion was summarized from these studies that the temperatures stratification realized in a given room and the rate of air exchange were not merely dependent on the room geometry and heat flux input, but also sensitive to the heat source geometry. More recently, Kaye and Hunt [2010] studied the effects of floor heat source area on the induced airflow in a room with natural displacement ventilation flows. They found that the temperature stratification in the room was very sensitive to the ratio of the heat source radius to room height. They developed a model to predict the transition process from displacement to mixing ventilation, which indicated that the transition was a combined function of the ratio of the heat source to the floor area, the heat source radius and the room height.

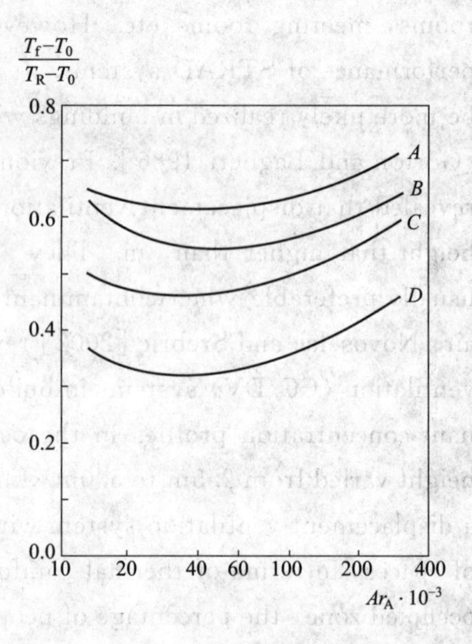

Figure 2.6 Minimum temperatures at floor level T_f versus Archimedes number for different, typical heat sources [Nielson, 1993]

A—Distributed heat source; *B*—Sedentary persons; *C*—Ceiling light; *D*—Point heat source

The locations of heat sources are not only significant to the contaminants concentration distribution as reviewed in section 2.2.1, but also greatly affect the temperature distributions in the room. Park and Holland [2001] used two-dimensional computational simulations to examine the effects of vertical location of a convective heat source on thermal DV systems. They found that the convective heat gain from the heat source to an occupied zone became less significant when elevating the location of the heat source above the floor. This effect changed the temperature field and resulted in the reduction of the cooling load in the occupied zone. In the REHVA Guidebook No. 1 [2002], the correlation between heat source location and gradient of dimensionless temperature ratio is presented, as shown in Figure 2.7. It indicated that when the heat sources were in the lower part of the room, the dimensionless temperature gradient in the lower part was larger than that in the upper part. When the heat sources were located mostly in the upper zone of the room, the temperature gradient in the lower part was smaller than that in the upper part.

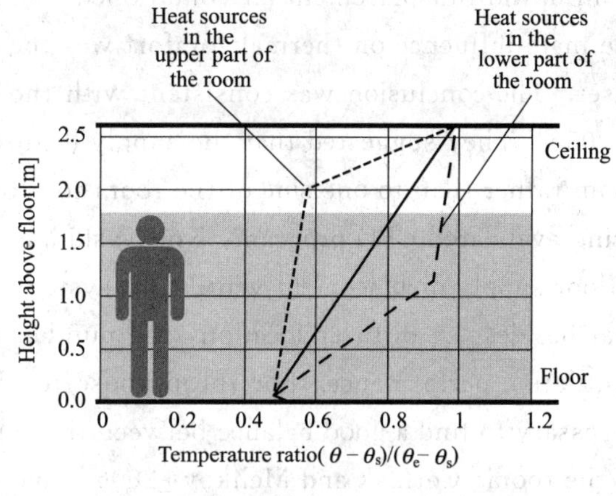

Figure 2.7 Temperature gradient in a displacement ventilated room with the heat sources at different levels [REHVA Guidebook No. 1, 2002]

Furthermore, the vertical temperature gradient is also dependent on the horizontal location of a heat source in a space. When a heat source is located close to a wall, the Coanda-effect will result in a reduction of air entrainment into the plume generated by the heat source. The flow in a plume closed to a wall was similar to one half of the flow in a free plume and had a convective heat emission twice than that of a single heat source [Nielson, 1993]. Thus, the stratification level in the space will shift upwards with a fixed flow rate of supply air. Previous studies indicated that the ventilation performance of a STRAD system was varied largely depending on the heat source location and type, for instance, the occupants, office equipment, and lights. However, the distribution of heat sources is mainly dependent on the function of room and it is too random to be predicted and systematically studied as a design parameter.

2.4.3 Supply diffusers' characteristics

The main difference between displacement ventilation (DV) and under floor air distribution (UFAD) systems is that the manners to deliver air into the space. The traditional DV system delivers air at very low velocities, while the UFAD systememploys higher velocity diffusers with correspondingly greater air mixing in the lower zone. As the terminal facilities of a STRAD system, the diffuser characteristics significantly affect the ventilation performance. It was described in REHVA Guide book No. 1 [2002] that "most draught problems reported from practical experience are due to the incorrect choice of diffuser". Yuan et al. [1998] conducted a parametric study on DV systems and they found that the diffuser types had a significant impact on the thermal comfort and energy performance of the systems. Lin and Linden [2005] investigated the influences of diffuser throw on thermal performance of UFAD systems in great detail. They concluded that the vertical momentum flux from the diffuser had significant effects on the temperature stratification distribution of an UFAD system. Lian and Wang [2002] studied thermal comfort in an upward displacement air conditioned room and the experimental results showed that the main influence on thermal comfort was the distance between the occupant and the diffuser. The conclusion was consistent with the simulation results presented by Lin et al. [2005c]. They suggested that the supply diffuser should be located near the center of the room rather than to one side of the room to provide a more uniform thermal condition. By using avalidated CFD program, Kobayashi and Chen [2003] evaluated the performance of a floor-supply displacement ventilation system with vortex diffusers. The results revealed that besides the diffuser location, the number of diffusers also had great impacts on the ventilation performance. For the purpose to obtain a better indoor environment, it was necessary to find a good balance between the supply air quality and the number of diffusers in the room. Cermak and Melikov [2006] investigated the air quality and thermal comfort with different airflow patterns in a climate chamber. They found that the UFAD system with a short vertical throw can ensure both a higher air quality and a lower risk of thermal discomfort as compared to that in the DV system or the UFAD system with a horizontal discharge pattern. Lee et al. [2009a, 2009b] systematically assessed the performance of various diffusers for different types of indoor spaces by using both experimental and numeric al approaches. The results showed that the linear diffuser created the highest velocity in the occupied zone, making the potential draft risk high, while the perforated-corner traditional DV diffusers and perforated-floor-panel diffusers would generate a high temperature difference between the head and ankle levels of an occupant. It could be concluded that the diffuser characteristics, including the type, throw height, location and number of diffusers, have great effects on the ventilation performance of STRAD systems and need to be taken into account in the design.

2.4.4 Layouts of exhaust and return grilles

In a well-mixed ventilation system, the location of exhaust grilles does not affect the space air-conditioning load significantly. However, due to the large vertical temperature stratification generated in STRAD systems, the locations of exhaust grilles play an important role in the energy performance of the systems.

As shown in Figure 2.8, Miyagawa [1983] revealed that the arrangement of air supply outlet and exhaust inlet greatly affected the heat load ratio contributed to the coil load, which was especially prominent for large space. Based on a series of experiments, Olivieri and Singh [1979] found that the locations of supply outlets and return openings played an important role on the thermal stratification of space air distribution and consequently influenced the cooling coil load. The lower the return air temperature, the less the capacity of the cooling unit required by the system. Gorton and Bagheri [1986] also studied the influence of return air inlets′ location on the cooling load. They supposed that the return air inlets should be located low, well within the cooled zone. As the height of return grilles varied from lower zone to upper zone, the cooling load increased and the

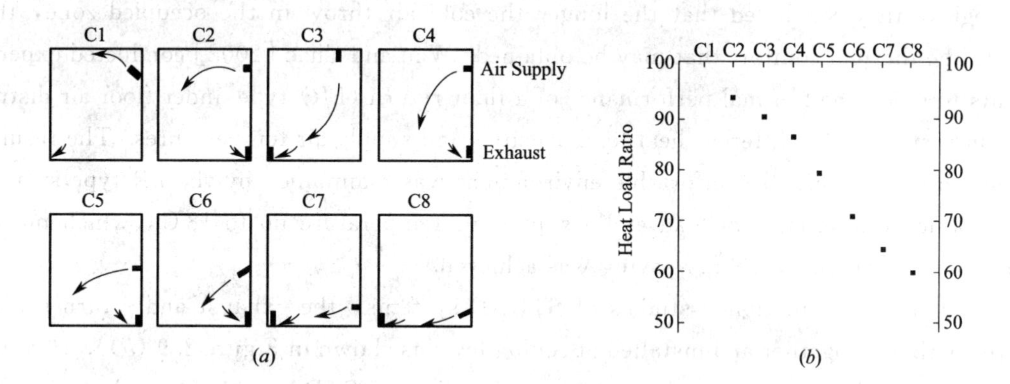

Figure 2.8 Effect of air supply and exhaust arrangement to the heat load ratio in large space
(*a*) Arrangement of air supply and exhaust vents in reference model; (*b*) Heat load ratio in cooling operation according to air supply/exhaust arrangement type [Miyagawa, 1983]

increment was dependent on building construction characteristics, load characteristics, and the upper zone temperature. Lam and Chan [2001] investigated the temperature distributions and air movement within an air-conditioned gymnasium in Hong Kong. The numerical results also revealed that the exhaust position had a great impact on the thermal stratification in the gymnasium and then significantly influenced the annual cooling load of the system. Similar with that, Awad et al. [2008] experimentally investigated the effect of extract terminal locations on the stratified flow characteristics in a ventilated chamber. It was found out that the exhaust location influenced the level of interface of the stratified layers, which consequently affected the cooling load of air conditioning units. Filler [2004] pointed out as well if return-air grilles were placed close to building's perimeter walls, convective heat gain occurring there could be removed directly by the return air before it entered into the occupied zone, which was benefit to save energy for the systems.

The locations of exhaust and return grilles not only greatly affect the energy efficiency of STRAD systems, but also play an important role on the air quality and thermal comfort of indoor environment. Li et al. [1999] numerically investigated the relationship between the air flow pattern and distribution of air age in a test chamber. They concluded that the location of return air inlet certainly influenced the distribution of air age, though the influence was not great. Holmberg and Chen [2003] worked on the air flow and particle distributions in a classroom with different ventilation systems to minimize respirable airborne particles in the breathing zone. The numerical results indicated that the exhaust conditions of the ventilation air flow had a great impact on the control of air quality. With an additional exhaust inlet below the breathing zone, high concentrations in the breathing zone can be controlled. Kuo and Chung [1999] studied the effects of the inlet and outlet diffuser's locations in the occupied region on the thermal comfort of occupants. They found with different locations of inlet and outlet diffusers, the distributions of thermal comfort parameters were different. By inspecting the simulation results from different ventilation strategies, they concluded that the longer the cold air throw in the occupied zone, the better the thermal comfort that may be obtained. Wan and Chao [2002] conducted experiments to study the thermal performance of a floor return (FR) type under-floor air distribution system under different heat load densities and supply air temperatures. The results indicated that a thermal comfortable environment was maintained by the FR type system under a dense internal heat load and a supply air temperature up to 18℃, which meant considerable amount of energy saving was achieved.

In most aforementioned studies of STRAD systems, the exhaust and return grilles were combined together and installed at ceiling level as shown in Figure 2. 9 (*b*). However, due to the large vertical temperature stratifications in STRAD systems and a quasi-upward directional flow in the upper space, local exhaust of higher temperature air directly to outside is possible to be realized and then the actual coil load of an air-handing unit (AHU) can be reduced. Based on onsite measurements and CFD simulation for an atrium of a demonstrative Integer Building in Hong Kong, Lau and Niu [2003] proposed that with separate locations of return and exhaust grilles, as shown in Figure 2. 9 (*d*), additional energy savings was able to be achieved for STRAD systems. By using CFD simulations, Xu et al. [2009] supposed that the extra energy saving potential for STRAD systems with similar configuration should be much more pronounced for large space with high ceiling.

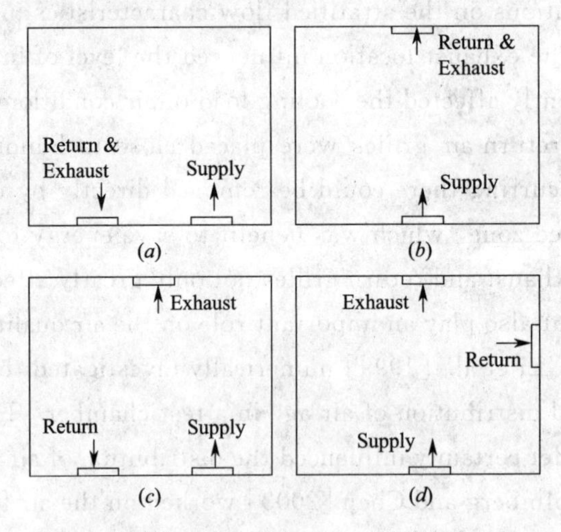

Figure 2. 9 Schematics of different combinations of UFAD systems [Xu et al. 2009]

2. 4. 5 Cooling load

2. 4. 5. 1 Cooling load diversity

Since the types and layouts of heat sources are varied largely among rooms with different functions, the cooling loads are substantially varied from project to project, which may greatly affect the indoor air quality and thermal comfort of STRAD systems [Bauman et al. 1991]. Liu and Hiraoka [1997] studied the impact of cooling load on the vertical temperature distribution with a fixed air change rate of 7 air changes per hour (ACH) for a UFAD system. The results revealed that the airflow patterns and the temperature distributions varied obviously with different cooling loads. Hagstrom et al. [2002] investigated the contaminant removal efficiency with different cooling loads in a room. The experimental results revealed that increasing the cooling load was able to improve contaminant removal efficiency clearly and form more uniform thermal environment inside the occupied zone, when locating the supply grilles at middle level of side walls.

Figure 2.10 presents some studies that investigated the range of cooling load per floor, in which acceptable thermal environments were able to be maintained in DV systems. To ensure that the maximum vertical temperature gradient in the occupied zone was not larger than 3K, Sandberg and Blomqvist [1989] suggested that the maximum convective cooling load in office buildings with traditional DV systems was not exceeded about $25W/m^2$. While other researches [Kegel and Schulz 1989, Svensson 1989] suggested that higher cooling load limits were affordable for DV systems. Chen and Glicksman [1999] also demonstrated that cooling load up to $120W/m^2$ could be handled in the office environment if increasing the ventilation rate. However, such airflow rates typically resulted in larger diffuser outlet areas in order to maintain the required low discharge velocities. Howe et al. [2003] reported that successful application of DV in a telecommunication equipment room with cooling loads up to $340W/m^2$. But thermal comfort did not need to be considered into in this application. It was concluded that when applied in a space with higher ceiling height, DV systems were capable of removing lager cooling loads [Skistad 1994]. By increasing the area of the air supply outlet or providing additional heat removal capacity, for instance, the installation of cooled ceiling, DV systems may also be applied to a space with a higher cooling load [Chen and Glicksman 2003].

2.4.5.2 Cooling load calculation methods

While the individual space heat gains encountered in STRAD systems differ little from those in mixed air systems, the cooling requirements of the space vary considerably. The existing cooling load calculation methods and the Radiant Time Series (RTS) method [ASHRAE 2009] are all based on the assumption that all the convective heat in the space contribute to the cooling load, which is obviously not suitable for STRAD systems. Therefore, modifications to current cooling load calculation methods are demanded to capture the thermal distribution characteristics in STRAD systems. In a STRAD system, the conditioned air is delivered directly to the occupied zone and thermal stratification is formed, which result in natural buoyant transfer of convective heat plumes. This means in STRAD systems, it is possible to prevent convective heat from upper zone being transferred to occupied zone. Bauman et al. [1995] conducted a series of experiments in a controlled environmental chamber with floor-based ventilation system. They concluded that by keeping the zone above the occupants head warmer, space-cooling energy could be reduced. Only the space heat gains that captured within the occupied zone or otherwise

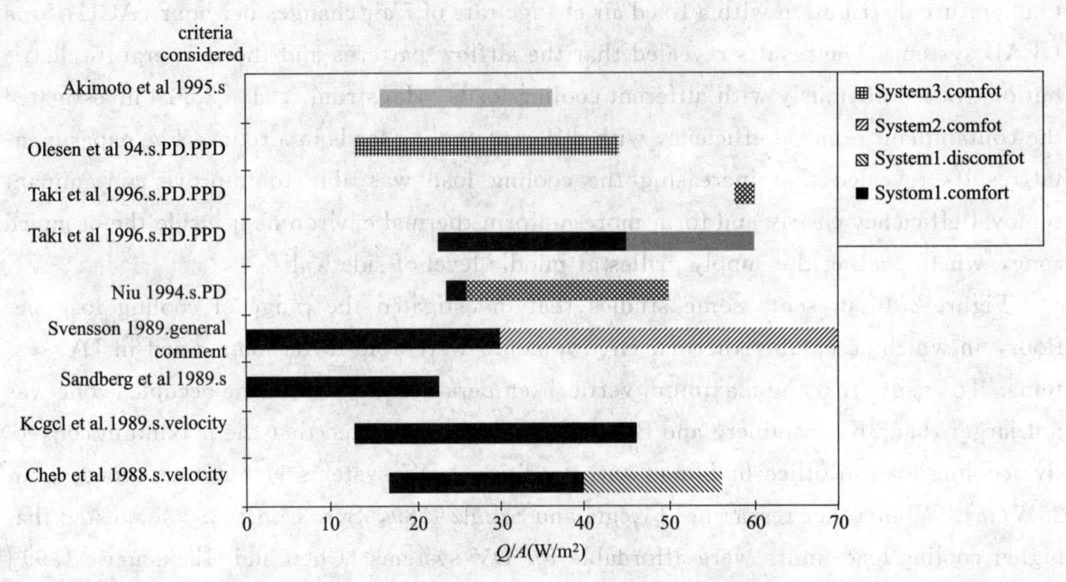

Figure 2.10 Ranges of cooling load per floor area for three types of displacement ventilation: side-wall diffuser (system 1), side-wall diffuser with cooled ceiling panel (system 2), and rise floor (system 3) [Chen and Glicksman, 2003]

affect the thermal comfort of occupants are need to be considered when calculating the required space supply airflow rate. Therefore, how to accurately calculate the portion of space cooling load that contributed to the occupied zone is crucial to determine the required supply air flow rate, which in turn significantly influences the ventilation performance.

The occupied zone in a space is any location where occupants normally reside in and should be carefully defined by designer. It usually considered as the lowest 1.8m of a room, although layers near the floor and walls are sometimes deducted from it. In REHVA Guidebook No. 1 [2002], the occupied zone is defined with details as shown in Figure 2.11, considering the postures of occupants. The height of the occupied zone is 1.3m for sedentary occupants and 1.8m for standing occupants.

Based on statistical data obtained from 56 displacement ventilation cases, Chen and Glicksman [2003] proposed a method to calculate the cooling load in the occupied zone and then determine the required supply airflow rate. This method can be applied to design the optimal combination of supply airflow rate and temperature to avoid large head-to-ankle temperature difference in the design stage. Loudermilk [1999] split the conditioned space into two distinct horizontal zones in UFAD systems, a mixing zone within the lower levels of the space and displacement type flow in the upper zone. He split the space sensible heat gains into convective and radiant components and supposed that the entire radiant gain must be considered in calculation of the supply airflow, while the convective heat gains that originate outside the occupied zone should be neglected. The effective heat gain factors (EHGFs) for individual space heat sources were defined to quantify their impact on the occupants in the space. The space effective sensible heat gain (ESHG) was obtained

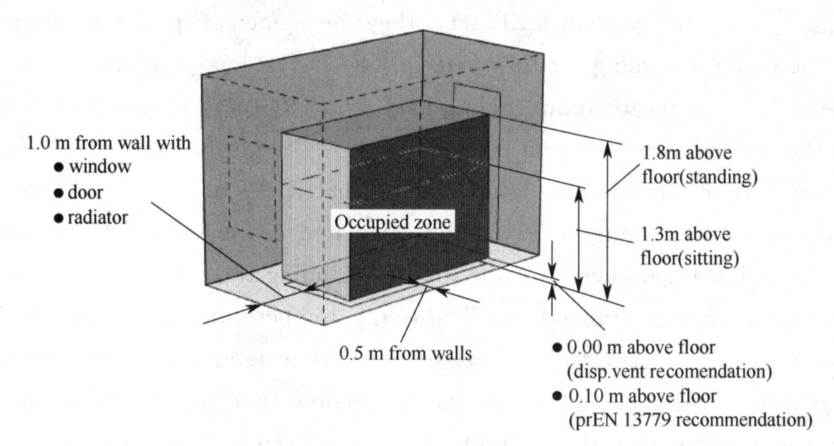

Figure 2.11 Definition of occupied zone [REHVA Guidebook No. 1, 2002]

based on the EHGFs and was used to determine the supply air flow rate. This method was based on the assumption that the height of the mixing zone is equal to the height at which the supply outlet discharge velocity has been reduced to 0.25m/s, i.e., throw height of diffuser. However, Liu and Linden [2006] theoretically and experimentally showed that the vertical projection of the outlet and the two layer stratification height are not equal, and the former cannot be used to predict the latter.

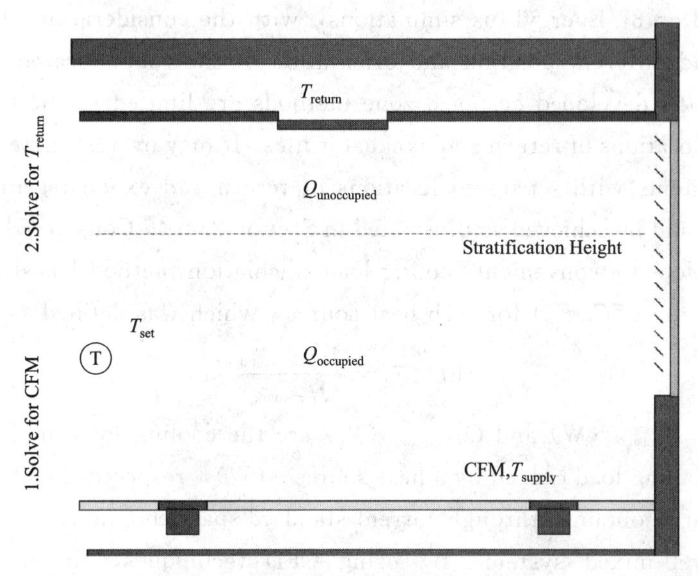

Figure 2.12 Definition of two zones for simplified UFAD load calculation model [Bauman, 2003]

A similar approach was subsequently proposed by Bauman [2003]. As shown in Figure 2.12, the space is separated into occupied zone and unoccupied zone. The stratification height was supposed to be equal to the upper boundary of occupied zone. Correspondingly, the space cooling load was artificially distributed into $Q_{occupied}$ and $Q_{unoccupied}$, and the former was used as the cooling load to calculate the supply air flow rate. However, this method was still at the conceptual level for information on how to obtain $Q_{occupied}$ was not provided. Later, they developed a practical design tool to determine the required cooling air flow rate in UFAD systems [Bauman et al. 2007]. A fixed value of 0.7 was set for room cooling load ratio (RCLR), which defined as the percentage of the total system heat gain (including 100% of lighting) that is to be assigned to the room in UFAD systems. The design tool predicted the room air temperature profile through empirical correlations based on the calculated cooling load that assigned to the room and the designed parameters of the room, including the room set-point temperature and the diffuser discharge temperature. The simple temperature profile was used to derive the two comfort parameters, the average occupied zone temperature and the head-ankle temperature difference, and then determine the airflow that matches most closely to the design conditions. Recently, the design tool was improved with much more flexibility and a simplified calculation method for designing cooling loads in UFAD systems was proposed [Schiavon et al. 2010]. The space cooling load in an UFAD system was calculated based on the cooling load ratio (UCLR), which defined as the ratio of the cooling load calculated for a UFAD system to that for a mixing ventilation system. The space cooling load was divided between the supply plenum, the zone and the return plenum, i. e. the supply plenum fraction (SPF), the zone fraction (ZF) and the return plenum fraction (RPF). Regression models to predict the UCLR and the split of the space cooling load have been developed based on 87 EnergyPlus simulations, with the consideration of different room floor levels, and different position and orientation of the room located. However, the applicability of the developed occupied zone methods are limited to the tested conditions with combined locations of return and exhaust grilles. It may overestimate the cooling load for STRAD systems with separated locations of return and exhaust grilles, since extra energy saving could be achieved as illustrated in Section 2.4.4. Considered for that, Xu et al. [2009] developed a convenient cooling load calculation method based on the effective cooling load factors ($ECLF_i$) for each heat source, which was defined as

$$ECLF_i = \frac{Q_{i-\text{occupied}}}{Q_{i-\text{space}}} \tag{2-7}$$

where, $Q_{i-\text{occupied}}$ (W) and $Q_{i-\text{space}}$ (W) are the cooling load in the occupied zone and the space cooling load caused by a heat source i (W), respectively. The space cooling load $Q_{i-\text{space}}$ can be obtained through current standard space cooling load calculation procedures for a well-mixed system. By using CFD techniques, the $Q_{i-\text{occupied}}$ can be determined according to the energy balance principles:

$$Q_{i-\text{occupied}} = H_{\text{r-plane}} - H_{\text{sup}} \tag{2-8}$$

$$H_{\text{r-plane}} = \sum (w_i A_i \rho C_p t_i)_{\text{r-plane}} \tag{2-9}$$

$$H_{\text{sup}} = V_s \rho C_p t_{\text{sup}} \tag{2-10}$$

where, $H_{\text{r-plane}}$ and H_{sup} are the enthalpy flow through the horizontal plane at the upper boundary of the occupied zone and the enthalpy flow of the supply air through all the supply diffusers. w_i (m/s), A_i (m^2) and t_i (℃) are vertical velocities, cross sectional areas and temperatures of the nodal points at the return level horizontal plane. $V_s(m^3/s)$ is the supply airflow rate and t_{sup}(℃) is the supply air temperature. In the calculation of $Q_{i-occupied}$, all other heat sources are turned off and the heat source i is the only heat source presented in the whole space. Similarly, the overall effective cooling factor $ECLF$ is defined as the following:

$$ECLF = \frac{Q_{occupied}}{Q_{space}} \tag{2-11}$$

where $Q_{occupied}$ (W) is the total space cooling load contribute to the occupied zone and Q_{space} (W) is the total space cooling load. Once the effective cooling load factors $ECLF_i$ and the space cooling load caused by each heat sources $Q_{i-space}$ are known, the total effective cooling load in the occupied zone $Q'_{occupied}$ originating from different indoor heat sources can be calculated conveniently as:

$$Q'_{occupied} = \sum_{i=0}^{n} ECLF_i O_{i-space} \tag{2-12}$$

The discrepancy between the total effective cooling load in the occupied zone $Q'_{occupied}$ and the total space cooling load contribute to the occupied zone $Q_{occupied}$ has been demonstrated to be less than 10% , which is acceptable for engineering design. However, this method is only validated in a small office with low ceiling level for simulated conditions and further in-depth investigations are required.

2.4.6 Supply air flow rate

The supply airflow rate is required to meet the thermal load in the space and crucial to the room air distribution. Holmberg et al. [1990] explored the effect of air flow rate on the inhaled air quality. They found that the benefit of improved indoor air quality for displacement ventilation (DV) systems was weaken when the air flow rate decreased to 5L/s per person, since the stratification height was far below the breathing height. Through using the age of air as a criterion, Xing et al. [2001] studied the impact of ventilation rate on the indoor air quality in a UFAD system. Their simulation results clearly showed that the higher the ventilation rate, the younger the age of the air, which meant that better IAQ was achieved. According to experimental results, Mundt [1995] stated that the temperature gradient in a DV system was significantly dependent on the ventilation flow rate and not greatly determined by the heat load distribution in the room. Tests were also conducted by Webster et al. [2002] to determine the impact of room airflow and supply air temperature on the thermal stratification in interior spaces with a UFAD system. The results revealed that spaces could be highly stratified when the airflow rate was reduced at a given load and the variation of supply air temperature had slightly effects on the temperature profile, which only moved to a higher or lower temperature. Nielsen et al. [2003] investigated the influence of varying supply airflow on thermal comfort parameters in a test chamber. They found that with increased supply airflow rate of a DV system, the predicted dissatisfied of occupants due to draft increased, while predicted dissatisfied of occupants due to vertical temperature gradient decreased. To satisfy the requirement listed in ASHRAE standard 55—2010 [2010] that fewer than 20% of people were dissatisfied, a very narrow range of acceptable flow rate existed for the examined case. In the design guide [2003], Bauman stated that increasing the airflow rate would raise the stratification height, thereby improve indoor air quality and reduce thermal stratification in the occupied zone. However, increasing the airflow rate required larger supply area to avoid thermal comfort problem due to local draft caused by too large supply air velocity. On the other hand, decreasing the airflow rate would reduce fan energy use. However, it would also reduce the stratification height and potentially decrease ventilation efficiency and thermal comfort in the occupied zone. Therefore, designing a reasonable supply air flow rate according to the occupied zone cooling load is crucial to the performance of STRAD systems.

2.4.7 Stratification height

In STRAD systems, the cold air is delivered directly to the occupied zone and normally returned at or near ceiling level. Thermal plumes developed around heat sources within the room entrain air from surrounding space and draw it upward. The upward movement of air in the room takes advantage of the natural buoyancy and produces a vertical temperature gradient. This buoyancy-driven floor-to-ceiling airflow pattern and vertical air temperature distribution are two main flow characteristics within spaces ventilated by STRAD systems. The stratification height (SH), which refers to the height in the room where the combined airflow rate of the thermal plumes equals to the total supply air volume entering the room, plays an important role in determining thermal, indoor air quality and energy performance. The stratification height typically divides the room into two zones with distinct characteristics. The lower zone predominantly contains cool fresh air and the upper zone contains warm, more polluted air. Thus, an important objective in designing and operating a STRAD system is to maintain the stratification height near the top of the occupied zone.

Sodec and Craig [1990] found out that the air temperature gradient in an indoor space with a STRAD system depended on the measuring point position, the cooling load, the ventilation rate and the diffuser shape. Based on an experimental study that focused on the stratification formed under idealized conditions, Hunt et al. [2001] pointed out that the controlling parameters on the thermal stratification were the buoyancy flux of the heat source, the volume flux and momentum flux of the cooling diffuser. In theoretical and experimental studies, Liu and Linden [2006] differentiated the throw height of outlet and the stratification height. They stated that the former could not be used to predict the latter. The throw height was a function of the balance between the initial momentum flux and the buoyancy flux from the outlet [Bloomfield and Kerr 2000], which depended on the airflow rate per diffuser, the outlet geometry and the temperature difference between the jet and the room. While the stratification height depended on the balance between the diffuser characteristics and the buoyancy plumes generated by the heat loads in the space. Experimental studies and numerical predications of the stratification in STRAD systems with different configurations, namely floor-return and top-return, were conducted by Wan and Chao [2005]. The results revealed that the temperature stratification in the enclosure was highly dependent on the thermal length scale of the floor supply jets and significant vertical temperature gradients occurred when the jet thermal length scale was ≤1. The thermal length scale was defined as the ratio of the buoyant jet length scale to the room height. The buoyant jet length scale for a buoyant jet was determined by the momentum flux and the buoyancy flux. Bauman [2003] also described that the stratification height was predominantly determined by the overall room air supply volume relative to the strength of heat sources in the space, and not by the vertical throw of the diffusers. Increasing the airflow rate or decreasing the cooling load would raise the stratification height and vice versa.

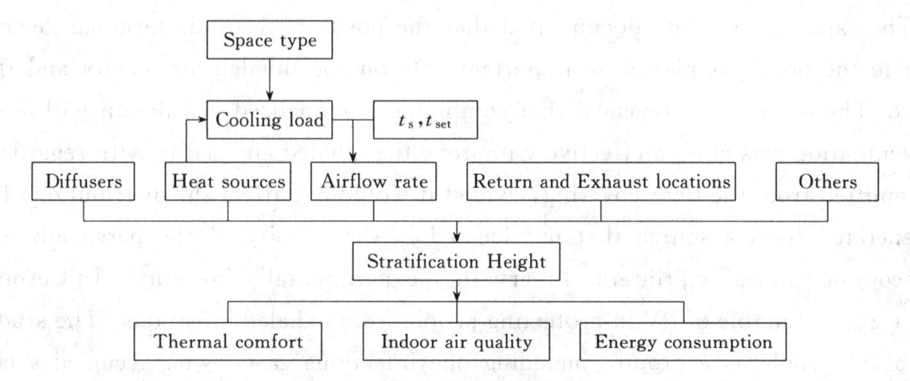

Figure 2.13 Key design parameters determining the stratification height

In a real building, other effects, for instance the wall characteristics and space height, would also influence the stratification and should be considered synthetically. It can be drawn from previous researches that the ventilation performance of a STRAD system is strongly affected by the stratification height, as presented in Figure 2.13, which in turn is mainly determined by several key design parameters, including the space type, the diffuser characteristics, the locations of return and exhaust grilles, the heat sources types and distributions, the designed airflow rate and so on. When designing a STRAD system, these parameters required to be considered synthetically to ensure the stratification height not lower than the breathing zone and to maximize the prospective advantages.

2.5 Integration of stratified air distribution system with personalized ventilation

Previous researches reported that using STRAD systems increased the likelihood of thermal comfort problems caused by local draft and excessive temperature stratification [Wyon and Sandberg 1990]. "Cold feet" caused by local draft and low supply air temperature was one of the main uncomfortable thermal sensations [Sekhar and Ching 2001]. Improving the supply air temperature to a warmer range is an effective way to avoid such uncomfortable sensation. However, it will lead to undesirable air temperature at breathing zone and complain of the "warm head" feelings. In this regard, recently the Personalized Ventilation (PV) was proposed to combine with STRAD systems in order to improve thermal comfort by cooling the facial region whilst keeping the feet region warm.

The PV system was proposed by Fanger [1999] for the purpose of supplying clean, cool and dry air to the breathing zone of occupants to achieve better inhaled air quality. The PV systems were recognized to have the ability to provide occupants with improved air quality, thermal comfort, individual control of the environment, and energy savings [Sekhar et al. 2005]. Several researches have been conducted to study the conjunction of PV systems with STRAD systems. Cermak et al. [2006] examined the performance of two efficient ways of supplying personalized air in conjunction with displacement ventilation. The experimental data documented that the position of an air terminal device with respect to the occupant played an important role on the inhaled air quality and thermal comfort. The results also revealed that combining personalized ventilation with displacement ventilation was able to effectively improve the inhaled air quality with regard to pollution emitted from the floor covering, whilst it would not affect the distribution of pollution generated from a source that not located in the vicinity of the personalized flow, e. g. , copy machines or printers. Later, they experimentally investigated [Cermak and Melikov 2007] the role of PV in protecting people from exhaled infectious. The study consisted of 30 people as a group, including one infectious case, who occupied a climatic room for eight hours. The results indicated that the pollutant concentration in inhaled air decreased clearly when PV was used for the exposed manikin. By using a numerical thermal manikin in the CFD simulations, Gao et al. [2006] also certified that coupling DV systems with PV was able to lower human exposure to ambient room pollutants and brings a "cool head" feeling that beneficial for the thermal comfort. Schiavon et al. [2007] stated that occupant's exposure to pollutant also depended on the ratio of time occupant stays at the workstation to the total time he/she stays in the room. They demonstrated that if the occupied density is lower than 0. 5, using displacement ventilation alone would be advantageous with regard to human-produced contaminates in comparison when it was combined with PV system. Based on an experimental study, Huang [2011] found that the air terminal devices played an important role on the performance of PV system when coupling it with a DV system. He recommended that the PV airflow rate should be kept at around 10～15L/s with a temperature of 21℃, when using a round movable panel (RMP) as the terminal device of PV system. From simulation results, He et al. [2011] concluded that applying RMP in DV and UFAD systems may sometimes increase the exposure of others to the exhaled droplets of tracer gas from the infected occupants for 0.8 mm particles and 5 mm particles, which was dependent on the personalized airflow rate. They supposed that PV could provide clean personalized airflow, but also increase the average concentration in the occupied zone of the exposed manikin. Whether the inhaled air quality could be improved or not by using PV was dependent on the balance of pros and cons of the system. Makhoul et al. [2012a] developed a room convective heat transfer model for DV systems assisted by PV systems. The effect of PV systems on the thermal plumes was considered. This model has been validated experimentally and used to estimate energy savings that resulted from using a higher supply air temperature in the DV system aided with a PV system. The potential for improving occupants' thermal comfort by coupling PV systems with UFAD systems was explored through a human response study [Li et al. 2010]. The results revealed that the acceptability of perceived air quality and thermal sensation were improved in integrated PV and UFAD system compared with that in UFAD system alone. The local thermal sensation at the feet was also improved when a warmer UFAD supply air temperature was adopted in the integrated PV-UFAD system. Supplying personalized air from front of the facial region provided cool feelings to the subjects' local thermal sensation and whole body sensation. Different with that, Makhoul et al. [2012b] examined the effect of assisting DV systems with PV on the segmental and overall comfort of the human body during transient load variations. The results indicated that an acceptable thermal comfort of 1.4 on a scale of -4 (very uncomfortable) to +4 (very comfortable) was obtained when a higher room air temperatures of the order of 26.8℃ was maintained in the occupied zone. The associated energy saving for a DV system aided by the personalized ventilator over nineteen hours of daily operation was about 27% of the energy used by a standalone DV system, while the same level of comfort was achieved. Halvonova and Melikov [2010a, 2010b, 2010c] explored the idea of using ductless personalized ventilators to aid with a DV system, by utilizing the lower clean air of the DV system. The impacts of disturbances due to walking persons and workstations layout, and the intake height on the performance of the integrated system were studied respectively. They reported that when maintaining an elevated room temperature, better thermal comfort and inhaled air quality was able to be achieved by supplying cool and fresh air at the breathing level of the occupants in a PV+DV system, compared to that in a standalone operating DV system. They suggested that a great distance should be kept between workstations with occupants seated back to back or a layout of workstations in which occupants were seated side by side.

Previous researches clearly demonstrated that combined a PV system with a STRAD system is beneficial for the thermal comfort of occupants, the inhaled air quality and the energy saving of the system. However, it is still a relative new pattern of indoor airflow. The interaction between the airflow driven by DV system and caused by PV system need to be further studied and a recognized design principle is also called for.

2.6 Brief review on the CFD simulation for indoor airflow

With the rapid advance in computer power and the development of user-friendly CFD program interfaces, CFD simulations have been becoming more and more popular in predicting ventilation performance in buildings. By solving a set of partial differential equations for the conservation of mass, momentum (Navier-Stokes equations), energy, chemical-species concentrations, and turbulence quantities, CFD simulations have the ability to present detailed information that concerning ventilation performance. Figure 2.14 presents the numbers of papers that using CFD to predict ventilation performance in buildings occupied by human beings from 2002 to 2007. It obviously indicated that the applications of CFD were mushrooming during the past few years.

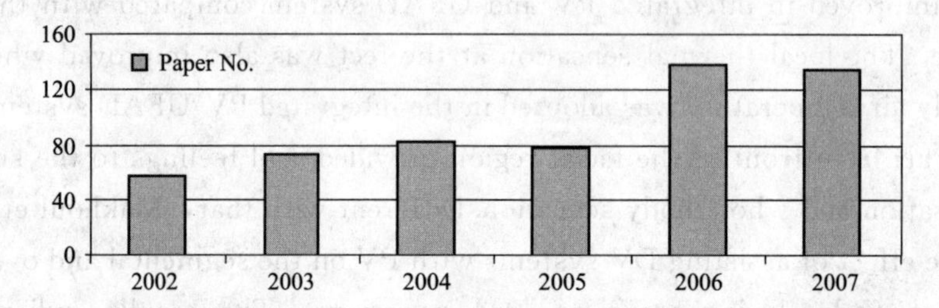

Figure 2.14 Paper numbers published from 2002 to 2007 in major journals using CFD to predict ventilation performance in buildings occupied by human beings [Chen 2009]

In REHVA Guidebook No. 10 [Nielson et al. 2007], a CFD simulation is described as a long and complex work process and consists of a large number of steps as shown in Figure 2.15. In each step, things may go wrong due to various reasons and lead to the prediction results greatly deviated from real situation to be modeled. For instance, the starting point for creating a CFD model of a building or a flow system is the representation of the building or system geometry as accurately as practical. It is very often to simplify the geometry for the CFD solution to enable efficient solution. However, unless the geometry simplifications are carried out very carefully and a sufficient understanding of the physical processes that can influence the flow, the final results may be affected. As presented in Figure 2.15, in the Guidebook the error sources are categorized into 7 groups with different properties. The first 4 groups refer to how good the model is set up, while Group 5 and 6 deal with the topic how good the equations are solved. Therefore, for the purpose of successfully using CFD models, a good knowledge of fluid dynamics and numerical technique is required for the users.

Since using CFD models does not necessarily ensure accurate results and involves engineering judgment, a procedure including verify, validate and report indoor environmental CFD analyses is required in a CFD simulation. Chen and Srebric [2002] have refined the definitions of these three steps as follows:

Errors	Uncertainties

Steps of CFD	Group
Define the problem	1. Define of problem
Define the geometry	
Generate a computation grid	2. Grid
Choose physical models	3. Models
Select turbulence model	
Define boundary conditions	4. Boundary conditions
Initial conditions	5. Numerical
Select a solution strategy	
Choose numerical procedure	
Solve the quequations	6. Code
Check the solution	7. User
Post-processing	
Analysis & interpretation	
Documentation	

Figure 2.15 Steps of CFD simulation and 7 groups of error sources

"The verification identifies the relevant physical phenomena for the indoor environmental analyses and provides a set of instructions on how to assess whether a particular CFD code has the capability to account for those physical phenomena.

The validation provides a set of instructions on how one can demonstrate the coupled ability of a user and a CFD code to accurately conduct representative indoor environmental simulations with which there are experimental data available.

The reporting of results provides a set of instructions on how to summarize the results from a CFD simulation in such a way that others who see the results can make informed assessments of the value and quality of the CFD work."

For STRAD systems, due to complicated air flow and distinct thermal stratification in the space, simple models, such as analytical and empirical models, have great uncertainties when used to predict the indoor environment. Thus, CFD models have been used as a convenient tool in such indoor environments to assess the ventilation performance [Chen 2009]. However, due to the approximations used in CFD models to make the transport equations solvable, the calculated results are inherently defective. In addition, CFD model with a validated turbulence model can be capable of visualizing the indoor environment of a space, but it may not suitable to apply for another space with different flow features. Thus, it is injudicious to suppose that CFD models could be used to replace all other models, such as the full-scale experimental model and the multi-zone network model [Chen et al. 2010]. Therefore, a reliable verification procedure is required for the applications of CFD technique in STRAD systems. As a convenient, economical and time saving tool, CFD models called for further efforts, including the development of more accurate and reliable turbulence models, and the improvement of numerical schemes.

2.7 Research gaps

The stratified air distribution (STRAD) system has been widely studied for the past few years. As a novel air-distribution method, the knowledge on STRAD is far from sufficient. Several research gaps are summarized based on foregoing reviews and presented as following:

ⅰ. Whether the occupant will accept the thermal stratification and feel comfortable or not is one of the main practical meanings in journey of pursuing massive applications of STRAD system. Several thermal comfort models that concerned on the non-uniform thermal environment in STRAD system have been developed. However, the evaluation results among different models may differ widely in their predictions [Jones, 2002]. Therefore, a detailed attention is required to be paid on the comparison of the evaluation results of these thermal comfort models, from view point of reliability and stability. Furthermore, the coupling procedures for most thermal comfort models with CFD simulations are manually required and complicated. Thus, how to simplify the coupling processes whilst ensure the evaluation accuracy is worth to be studied.

ⅱ. STRAD systems with separate locations of return and exhaust grilles have been proposed to be possible to achieve further energy savings [Lau and Niu 2003, Xu et al. 2009]. However, the study is still at preliminary stage and limited to the simulation conditions. Therefore, an experimental study focused on the performance of STRAD systems with separate locations of return and exhaust grilles is called for. In addition, the thermal flow characteristics in STRAD systems with similar configuration are required to be further illustrated. Furthermore, for STRAD systems with separate locations of return and exhaust grilles, there are rare investigations on the energy saving principle of cooling coil and the corresponding cooling coil load calculation method. Therefore, a reliable method that able to calculate the cooling coil load in STRAD systems is called for to be developed and then used to determine the equipment sizing properly.

ⅲ. A novel CFD-based occupied zone cooling load calculation method was developed by Xu et al. [2009]. However, the method is only validated in a small office with low ceiling level for tested conditions. Therefore, further validations of this occupied zone cooling load calculation method for the applications of STRAD systems in different space types are also required. Furthermore, a database that contained the effective cooling load factors for different heat sources in different spaces is demanded to be built up. The database then can be conveniently adopted by consultant engineers to calculate the occupied zone cooling load in a space and determine the required supply air flow rate, which is crucial to the thermal stratification and ventilation performance of STRAD systems.

ⅳ. The diffusers' locations have been demonstrated that strongly influence the thermal stratification and energy efficiency of STRAD systems. The influence is supposed to be much more prominent in STRAD systems with separate locations of return and exhaust grilles. Therefore, the influences of different locations of return and exhaust grilles on the performance of STRAD systems with such innovative construction, in terms of thermal comfort and energy saving, are also need to be examined in detail for different space types.

2.8 Summary

This Chapter gives a systematic review on STRAD systems, including the advantages and evaluation criteria of ventilation performances. Several key design parameters that significantly influence the performances of STRAD systems were also reviewed in detail. A well-designed STRAD system is recognized for its better ventilation performance in comparison of that for a well-mixed ventilation system. However, the much more complicated air flow in a STRAD system is demand more rigorous analysis of thermal and airflow processes in the space, which may be beyond the capability of most practicing consulting engineers. In addition, further efforts are required to make on the occupied zone cooling load calculation methods, and the corresponding reliable cooling coil load calculation methods.

This book concerns on the thermal comfort and energy saving for STRAD systems, especially for the ones with separate locations of return and exhaust grilles. Two representative thermal comfort models are investigated in detail when used to evaluate the thermal stratified environments, and the evaluation results are compared. New coupling procedure between a comprehensive thermal comfort model and CFD simulations is proposed. By splitting the locations of return and exhaust grilles in a STRAD system, extra energy saving potential is able to be provided. The energy saving of the cooling coil is associated with but differential from the space cooling load reduction. A novel CFD-based cooling coil load calculation method that can be used to evaluate the energy saving and determine the cooling coil load of a STRAD system is developed. Experiments are carried out to illustrate the energy saving principles for STRAD systems, and to validate the developed cooling coil load calculation method. Numerical studies are conducted as well to illustrate the thermal and airflow processes, when applying STRAD systems in three common types of space in Hong Kong. The impacts of diffusers' locations on the performance of STRAD system are investigated. A database is built up, which contains the cooling load factors for different heat sources in different building space types. The database can be adopted by engineers to calculate the occupied zone cooling load and then determine the required supply airflow rate. The main aim of this book is to provide a reference to engineers for the properly design of STRAD systems, including the required air flow rate determination, the occupied zone cooling load calculation, the cooling coil load calculation and the diffusers locations.

Chapter 3 Comparisons of two typical thermal comfort models

3.1 Introduction

Due to asymmetrical thermal environments are formed in STRAD systems, the most frequently cited thermal comfort standards: ASHRAE 55 [2010] and ISO 7730 [1994] may not be able to evaluate thermal comfort in such conditions with regard to the overall efforts combined with asymmetrical radiation and local airflow. Therefore, as reviewed in Section 2.3.2, extensive investigations focused on thermal comfort evaluation in inhomogeneous environments have been conducted and a number of thermal comfort evaluation models have been developed during past decades. However, there are rare comparisons among these thermal comfort models, which may differ widely in their predictions. In addition, the characteristic and applicability of different thermal comfort models also needed to be clarified. Therefore, in this chapter, the thermal environments in a small office ventilated by STRAD systems are going to be assessed by using two typical models, the simple ISO standard 14505 [2007] method and the comprehensive UCB thermal comfort model [Zhang 2003]. The coupling procedures of these two models with Computational Fluid Dynamic (CFD) simulations and the thermal comfort evaluation results will be compared and discussed in details.

3.2 Different coupling procedures

The UC Berkeley thermal comfort model (UCB model) is recognized for its excellent performances in the predictions, but manual coupling procedure with CFD simulation is needed. Gao et al. [2007] coupled the UCB model with a CFD simulation to evaluate human thermal comfort in a displacement ventilation system integrated with personalized ventilation (PV). Figure 3.1(*a*) is the schematic diagram of the calculation process with an iteration loop. First, the environmental parameters around human body, including the velocity and temperature, are calculated in CFD simulations based on initial boundary conditions; then manually input these parameters into the thermoregulation module of the UCB model to obtain the new skin surface temperatures. If the iteration is not convergent, the new skin surface temperatures will be imported to CFD as the new boundary conditions for the next cycle. The coupling process will be a bit of inconvenience if several iteration loops are needed to get convergence since it cannot be conducted automatically.

Different from the UCB model, the ISO standard 14505 [2007] is public and convenient to be adopted. The flow chart of the calculation procedure when using the ISO Standard 14505 method is displayed in Figure 3.1(*b*). The human body surface is fixed with a constant skin temperature and the effect of clothing is considered by adding a resistance to the skin surface. Only the sensible heat flow of each segment is need to be output from the CFD simulation and used to calculate the equivalent temperatures. The coupling procedure is obviously simpler than that in using the UCB model. However, the reliability of this method is undefined and needed to be further validated.

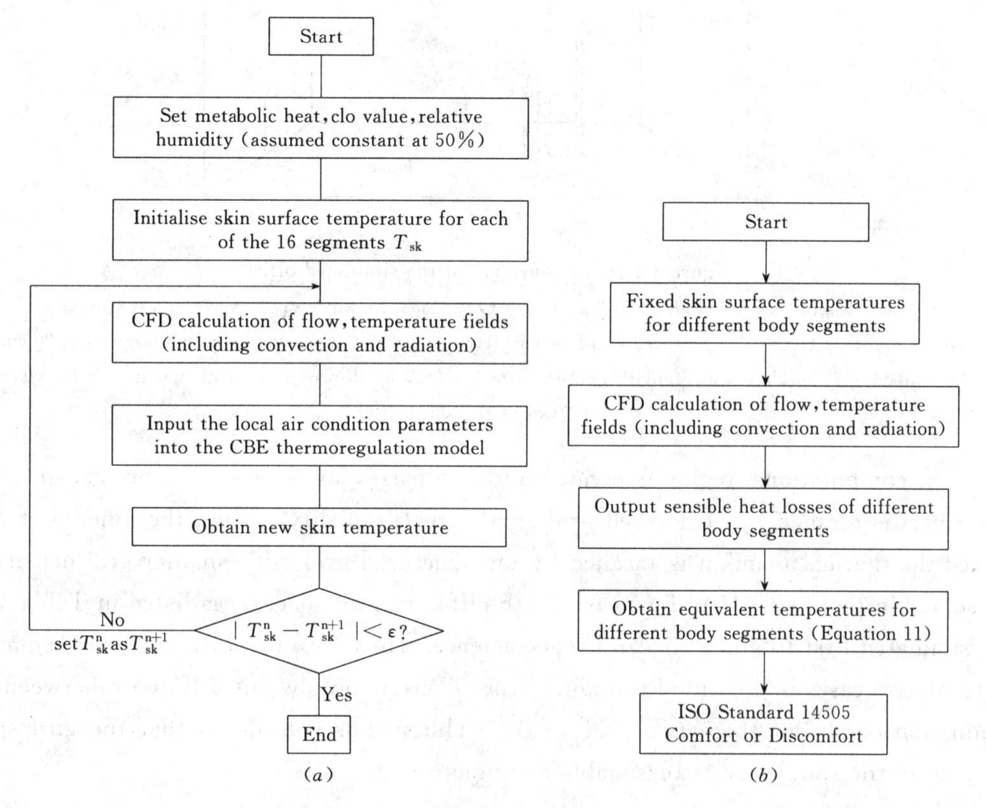

Figure 3.1 Evaluation methods of different thermal comfort models

(*a*) Coupling of CFD and the UC Berkeley thermal comfort model [Gao et al. 2007];

(*b*) Flow chart of the ISO Standard 14505 method with CFD

3.3 CFD simulations

Thermal environment in a small office ventilated by a displacement ventilation system aided with a PV system, which was presented by Gao et al. [2007] and shown in Figure 3.2, is going to be re-evaluated by using both the UCB model and ISO Standard 14505 method. The supply air temperature for DV system is constant at 19℃, which is set to 20℃ for PV system. The settings for the CFD simulations are same to that as described by Gao et al. [2007], including a space heat load of 37W/m^2 and a constant total supply airflow rate of 51L/s. The conduction heat through clothing is considered in present study as heat transfer through thin shell with insulation level of 0.5clo, which is the main difference with that done by Gao et al. [2007].

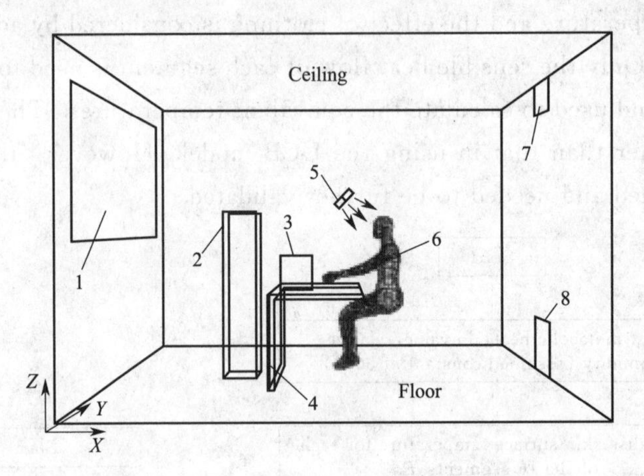

Figure 3.2 Configuration of the simulated office

(room length (X) 4m, width (Y) 3m, height (Z) 2.7m; 1—window; 2—vertical heat source; 3—computer; 4—table; 5—personalized ventilation air terminal device (circular outlet with a diameter of 20 cm); 6—human body; 7—displacement ventilation outlet 0.4m ×0.3m; 8—displacement ventilation inlet 0.4m×0.5m) [Gao et al. 2007]

The computational region is divided into two parts, as shown in Figure 3.3(*a*). The outer part is meshed by structured grids with greater quality, while the inner part contained the thermal manikin is meshed by un-structured grid with smaller grid density as presented in Figure 3.3(*b*). Two cases with different grid spaces, as listed in Table 3.1, are simulated first to study the grid independence. Heat flux for different body segments in these two cases is presented in Figure 3.4. There is no obvious difference between the simulation results of these two grid spaces. Thus, it is considered that the gird space adopted in the simulation is reasonable and practicable.

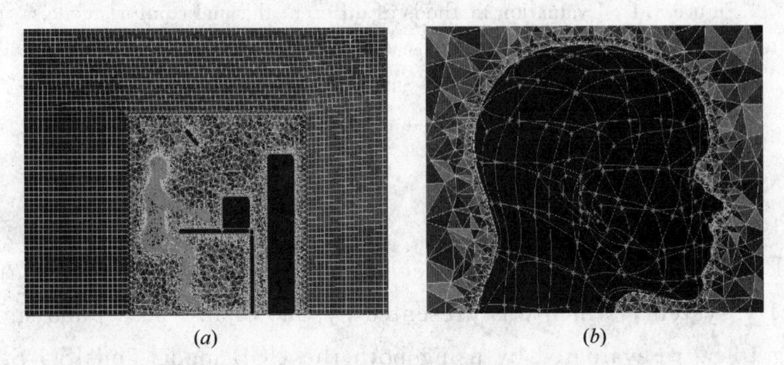

(*a*) (*b*)

Figure 3.3 Cross-sectional views of the grids system

(*a*) Mesh generation in the simulated office; (*b*) The grids surrounding the head

Two cases with different grid spaces **Table 3. 1**

Grid Space (mm)	Inner	Outer	Total-Elements
Case 1	20	50	1742618
Case 2	10	30	3280934

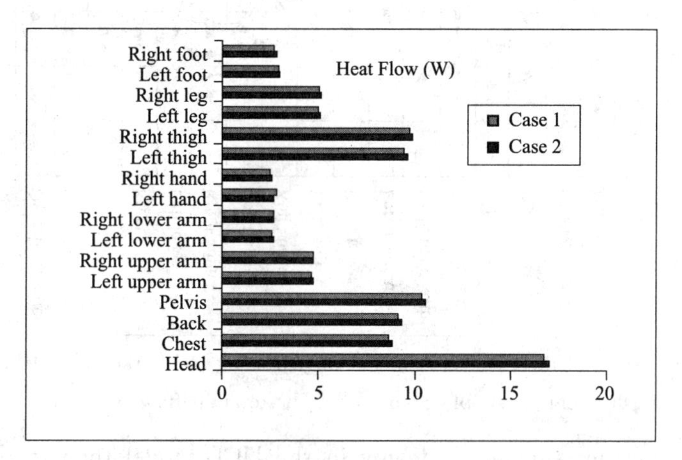

Figure 3. 4 Heat flux for different body segments

3. 4 Comparison of thermal comfort evaluation results

Three typical conditions, corresponding to the settings of PV air flow rate at 0L/s, 10L/s and 15L/s respectively, were simulated. When the air flow rate of PV system is set to 0L/s, it means that a pure DV system is used. Figure 3. 5 presents the thermal sensations of human body segments under different air supply conditions, evaluated by using ISO Standard 14505 method and UCB model respectively.

The overall thermal sensation predicted by UCB model was close to neutral situation. When the space was ventilated by DV only with no PV, the overall sensations predicted by both models were located at cool side. As reviewed before, for the occupants staying in a room that ventilated by a pure DV system, it is frequently to have "cold feet and warm head" feeling, which lead to thermal discomfort. This has been validated by the prediction results of UCB model, but the ISO method obtained contradictory results and indicated the head "felt" colder than the feet. The main reason was that in CFD simulation, the heat loss of head was more than the physical truth due to ignoring the thermal resistance of hair. Therefore, "cold head" feeling was predicted by ISO method, since it used the sensible heat loss of head directly to predict the thermal sensation, while the thermal resistance of hair was considered in the physiological module of UCB model. If servicing personalized air at 20℃ from 0L/s to 15L/s, the overall sensations predicted by both models were shifted to cooler side, but more sensitive when using the ISO Standard 14505 method. However, the sensations of those body parts which were directly exposed to personal-

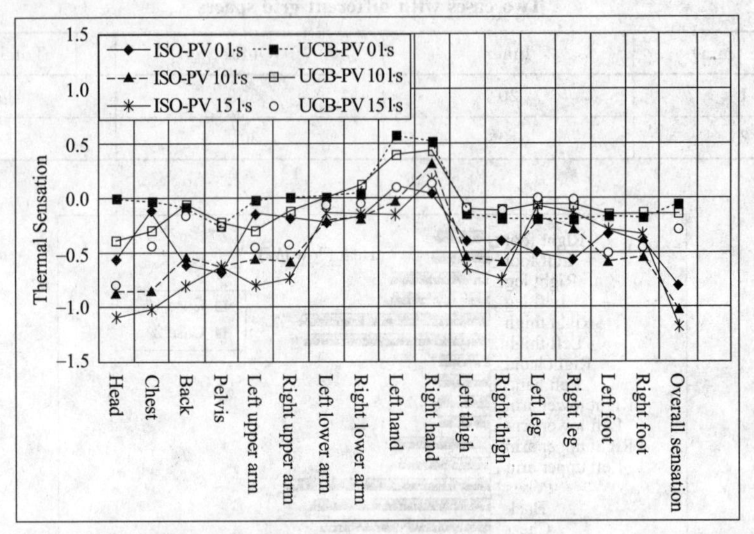

Figure 3. 5 Comparison of thermal sensations evaluated using UC Berkeley model and ISO Standard 14505 model with different air supply conditions (The total air flow rate is a constant at 50L/s)

ized air shifted to cooler side more deeply in the UCB model than in the ISO Standard 14505 model. For example, sensations of the head decreased numerically by about -0. 8 in the UCB model and -0. 5 in the ISO Standard 14505 method respectively. Thus, it is suggested that in an overall thermally neutral environment, the ISO Standard 14505 method is less sensitive than UCB model to local draft when predicting the local thermal sensations for the cooling variation. Figure 3. 3 also reveals that the thermal comforts of lower body parts are beneficial from the increasing of personalized air flow rate, which is consistent with the findings concluded by Li et al. [2010]. But the degree of such benefit predicted by these two methods were some different. For instance, the sensation at the legs improved numerically by around 0. 2 in UCB model and about 0. 5 in the ISO model respectively, when increasing the flow rate of PV system from 0L/s to 15L/s. Therefore, it is considered that in an overall thermally neutral environment, the ISO Standard 14505 method is more susceptive than the UCB model to the warming variation. For the same environment, both the local and overall thermal sensations evaluated by ISO Standard 14505 method were slightly "cooler" than that evaluated by UCB model.

3. 5 Discussion and Summary

Human thermal sensation and comfort in asymmetry environments are complicated physiological and psychological responses. Investigations on this topic have been conducted for decades and several excellent models have been promoted by a number of researchers. However, it was observed that most of them were not comprehensive enough or limited to specific environments, and only a few of them addressed human responses to both non-uniform and transient conditions with detailed thermoregulation model. The UC

Berkeley thermal comfort model was developed in recent years and has been validated by a large number of experimental tests. The model considered the physiological interactions between the local body parts and the whole body thermal sensation, and successfully built the calculation model of the local and whole body thermal sensation and comfort for asymmetrical conditions. But the coupling procedure between the model and CFD simulation is complex and an iteration loop is required, as presented in Figure 3.1. This is the main reason, for the time being, limit the possibility of widely spread of the model. In addition, as highlighted in Figure 3.4, the air temperature and velocity surrounding human body, and the surface temperatures of building envelopes obtained in CFD simulations are required to output to the physiological module and calculate the convective and radiation heat loss of human body. Then, the skin surface temperature and the core temperature of human body predicted by physiological module are used to evaluate thermal comfort of occupant in psychological module. This procedure can run smoothly for a space with simple construction. However, when assessing the thermal environment in a space with complex structure, such as the lecture theatre with terraced floor, it is difficult to re-built the geometry in the physiological module and reappear the temperature distributions of the enclosure surfaces. In addition, since the surface temperatures of enclosures are also stratified in STRAD systems, it is hard to represent the surface temperature distributions in the physiological module. Thus, when calculating the operative temperature and radiation heat transfer between human body and environment, simplified and/or approximate approach is adopted, which will inevitably lead to additional computational error. Therefore, it is envisaged that in the long run, as indicated in Figure 3.6 by arrows with dash line, practically reliable prediction of heat transfer between the human body and environment can be obtained with CFD methods, which can be fed back directly to the human body thermal regulation model so that the thermal sensation and comfort of different body segments can be evaluated more conveniently and accurately.

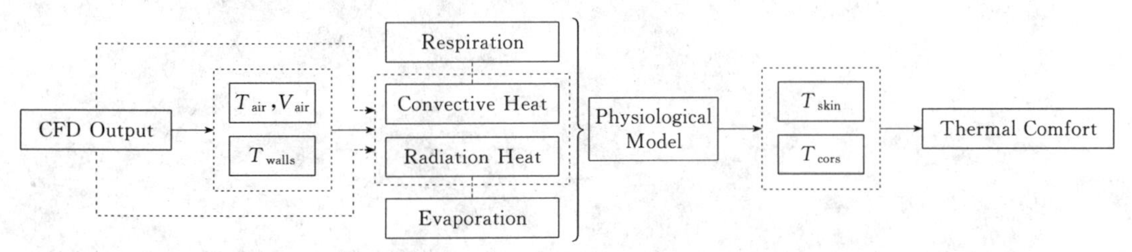

Figure 3.6 Coupling procedure of UCB thermal comfort model with CFD simulation

On the other hand, the ISO Standard 14505 method adopted the equivalent temperature as an index to evaluate the thermal sensations in asymmetrical environments conveniently. The thermal sensation, primarily due to local sensible heat variations, is evaluated by using the clothing-independent thermal comfort diagrams. But as a convenient way, this method does not include the thermoregulation model and the reliability of the method is required to be treated with caution when evaporation from skin is involved. According to our case study, in a thermal neutral environment, the ISO Standard 14505 method is more sensitive to the warming variation, while less sensitive to the cooling variation, compared with the UC Berkeley thermal comfort model. This may imply that the ISO Standard 14505 is just suitable for thermal neutral situation, where the latent heat of evaporation constitutes a very small part of the total heat transfer of human body. The model revealed the facial cooling effect of personalized ventilation well, which is beneficial to the thermal comfort of occupants in thermal stratified environment and in agreement with the sophisticated UCB model predictions. But the thermal resistance of hair is required to take into account to avoid the false feeling of "cold head" in the prediction results, when using the ISO Standard 14505 method to assess stratified thermal environments.

Chapter 4 Cooling coil load calculation method for stratified air distribution systemsand Experimental validation

4.1 Introduction

A novel CFD-based occupied zone cooling load calculation method in STRAD systems has been developed by Xu et al. [2009]. They also claimed that by splitting the locations of return and exhaust grilles, extra energy saving potential was provided. However, statistical validation and reliable evaluation method for the additional energy saving are still lacking. Furthermore, there is no direct investigation on the cooling coil load calculation method for STRAD systems with separate locations of return and exhaust grilles. Therefore, in this chapter, the energy saving principle for STRAD systems with separate locations of return and exhaust grilles is theoretically clarified first. Then, a novel cooling coil load calculation method for STRAD systems is developed. For the first time, experimental study on a STRAD system with separate locations of return and exhaust grilles is also conducted in a full-scale climate chamber. The additional energy saving for STRAD systems obtained by splitting the locations of return and exhaust grilles is clarified by using the experimental results. The newly developed cooling coil load calculation method is demonstrated that it is feasible to be used for STRAD systems with separate locations of return and exhaust grilles.

4.2 Energy saving and Cooling coil load calculation

The STRAD systems are inherently energy-saving since only the lower occupied zone is required to be cooled. The typical air flow distribution in a well-designed STRAD system is presented in Figure 4.1.

The return and exhaust grilles are combined together and located at suspended ceiling. The stratification height is higher than the upper boundary of the occupied zone and there is no reversed flow from upper unoccupied zone to lower occupied zone. Thus, the convective heat that generated in the upper unoccupied and released to the exhaust/return air is able to be discharged from the space directly and do not contribute to the occupied zone cooling load. However, due to the combined locations of return and exhaust grilles, the return and exhaust air is mixed together and warmed up simultaneously. Thus, portion of the

Figure 4. 1 Typical air flow in a well-designed stratified air distribution system with combined return and exhaust grilles

convection heat gains is still need to be afforded by the cooling coil load of air handling unit (AHU).

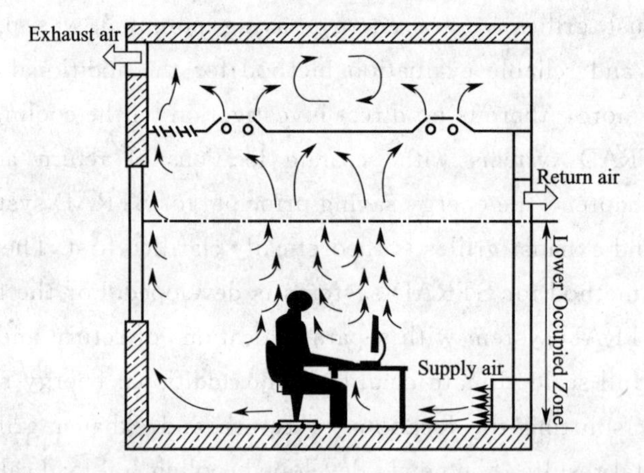

Figure 4. 2 A stratified air distribution system with separate locations of return and exhaust grilles

Different with that, by spitting the locations of return and exhaust grilles, further energy saving potential is provided. As illustrated in Figure 4. 2, in a STRAD system with exhaust air at ceiling level and return air at middle level, it is possible to impel much more convective heat, especially those generated at the upper zone, from exhaust grille directly. Therefore, even the occupied zone cooling load in a STRAD system with combined locations of return and exhaust grilles may be same to that in a STRAD system with separate locations of return and exhaust grilles, the energy saving of the cooling coil in the latteris more prominent than that in the former.

How to exactly evaluate the energy saving of the cooling coil and calculate the cooling coil load is important for the design of STRAD systems. As presented in Figure 4. 3 (a), in a well-mixed ventilation system thecooling coil load Q_{coil} (W) of the air handling unit (AHU) is equal to:

$$Q_{\text{coil}}=C_{\text{p}}\times\dot{m}_{\text{r}}\times(t_{\text{r}}-t_{\text{s}})+C_{\text{p}}\times\dot{m}_{\text{o}}\times(t_{\text{o}}-t_{\text{s}}) \tag{4-1}$$

where, $\dot{m}_{\text{r}}$ (kg/s) is the return air flow rate and t_{r}(℃) is the return air temperature.

$\dot{m}_o$(kg/s) is the outdoor air flow rate and t_o(℃) is the outdoor air temperature. C_p[J/(kg·K)] is the air specific heat capacity and t_s(℃) is the supply air temperature. For the air flow balance in the room, the outdoor air flow rate $\dot{m}_o$ is equal to the exhaust air flow rate $\dot{m}_e$(kg/s). Therefore, with some further algebra operations, Equation (4-1) can be written as.

$$Q_{coil}=C_p\times\dot{m}_r\times(t_r-t_s)+C_p\times\dot{m}_e\times(t_e-t_s)+C_p\times\dot{m}_o\times(t_o-t_e) \quad (4\text{-}2)$$

where, t_e(℃) is the exhaust air temperature. Since the air mixed well in the room, the exhaust air temperature is equal to the room set-point temperature t_{set}, which is also equal to the return air temperature, i.e. $t_r=t_e=t_{set}$. Thus, Equation (4-2) can be written as:

$$Q_{coil}=C_p\times\dot{m}_r\times(t_r-t_s)+C_p\times\dot{m}_e\times(t_e-t_s)+C_p\times\dot{m}_o\times(t_o-t_{set}) \quad (4\text{-}3)$$

The space cooling load Q_{Space} (W) and the ventilation load Q_{vent} (W) in the classical sense can be calculated as:

$$Q_{Space}=C_p\times\dot{m}_r\times(t_r-t_s)+C_p\times\dot{m}_e\times(t_e-t_s) \quad (4\text{-}4)$$

$$Q_{vent}=C_p\times\dot{m}_o\times(t_o-t_{set}) \quad (4\text{-}5)$$

Substitute Equation (4-4) and Equation (4-5) into Equation (4-3), the following equation can be obtained:

$$Q_{coil}=Q_{Space}+Q_{vent} \quad (4\text{-}6)$$

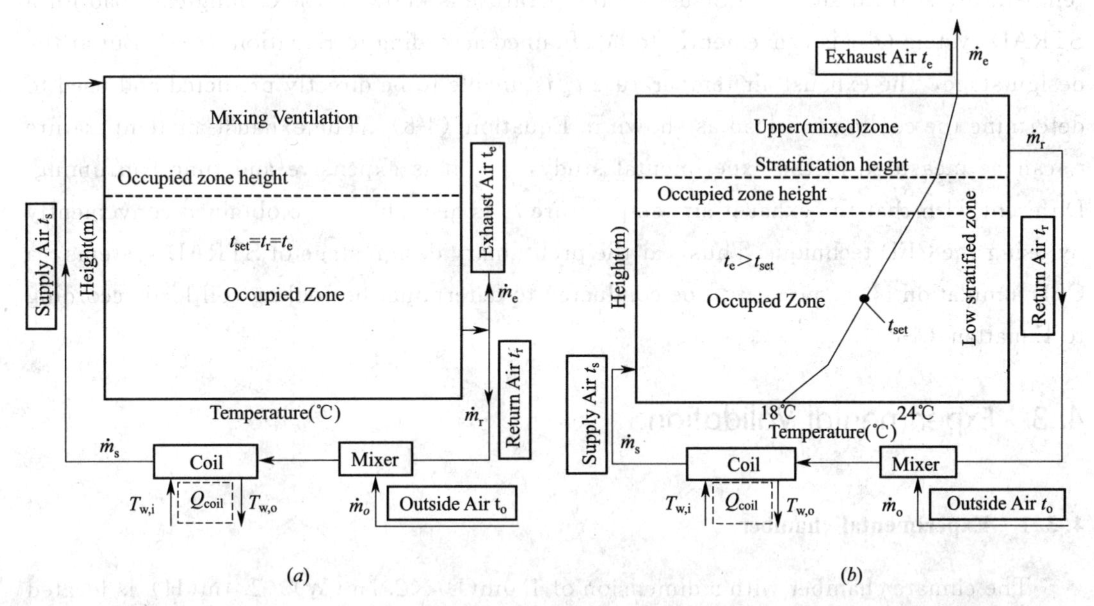

Figure 4.3 Sketch maps of different air distribution methods

(*a*) Mixing ventilation system; (*b*) Stratified air distribution system with separated location of return and exhaust grilles

For a STRAD system as shown in Figure 4.3 (*b*), the cooling coil load of AHU can stillexpressed as Equation (4-1), and Equation (4-4) and Equation (4-5) are effective as well. What notable is that due to thermal stratification in the space, the exhaust air temperature t_e is no longer equal to the set-point temperature t_{set} and the former is much larger than the latter. Equation (4-4) can be rearranged as

$$C_p \times \dot{m}_r \times (t_r - t_s) = Q_{Space} - C_p \times \dot{m}_e \times (t_e - t_s) \quad (4\text{-}7)$$

Substitute $C_p \times \dot{m}_r \times (t_r - t_s)$ in Equation (4-7) with Equation (4-1), and with some further algebra operations we have

$$Q_{coil} = Q_{Space} + Q_{vent} - C_p \times \dot{m}_e \times (t_e - t_{set}) \quad (4\text{-}8)$$

Compared Equation (4-8) with Equation (4-6), it is obvious that Q_{coil} in a STRAD system is smaller than that in a well-mixed ventilation system by the quantity $C_p \dot{m}_e (t_e - t_{set})$. In other words, if we define as ΔQ_{coil}

$$\Delta Q_{coil} = C_p \dot{m}_e (t_e - t_{set}) \quad (4\text{-}9)$$

then, ΔQ_{coil} is the energy saving of the cooling coil of STRAD systems. It indicated that the reduced AHU cooling load in STRAD systems is reversely related to the exhaust air temperature. It should be noted that Equation (4-8) and Equation (4-9) hold true to both split and combined exhaust/return designs. By splitting the return and exhaust grilles, the exhaust air temperature may further raise up, which means much more heat is discharged to the outside directly, and therefore extra energy savings in STRAD system is achieved. The space cooling load Q_{Space} and the ventilation cooling load Q_{vent} can be calculated according to ASHRAE Standard [2009]. Therefore once the room set-point temperature is fixed and the exhaust air temperature is known, the cooling coil load for a STRAD system Q_{coil} is conveniently to be obtained according to Equation (4-8). But at the design stage, the exhaust air temperature t_e is unable to be directly predicted and used to determine the cooling coil load as shown in Equation (4-8) . The exhaust air temperature t_e can be measured in the experimental study, but it is expensive and time consuming. Different with that the exhaust air temperature t_e is also able to be obtained conveniently by using the CFD technique. Thus, at the preliminary design stage of STRAD systems, a CFD simulation is encouraged to be conducted to determine the cooling coil load according to Equation (4-8) .

4.3 Experimental Validation

4.3.1 Experimental chamber

The climate chamber with a dimension of 4.0m(L)×2.7m(W)×2.4m(H) is located in a laboratory of The Hong Kong Polytechnic University as shown in Figure 4.4. The enclosures of the climate chamber are sealed well with thermal insulation materials.

The laboratory is conditioned by the mixing ventilation system of the building, while the climate chamber is ventilated by an independent STRAD system as indicated in Figure 4.5. All of the air is supplied from floor level with cool temperature. It should be noted

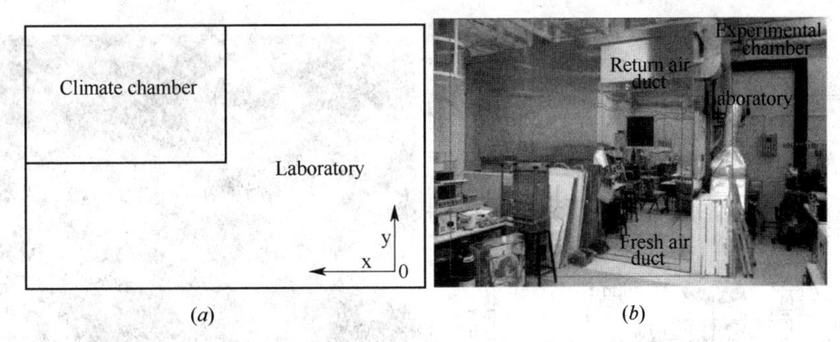

(a) (b)

Figure 4. 4 Sketch map and reality image of the climate chamber

(a) Sketch map; (b) Reality image

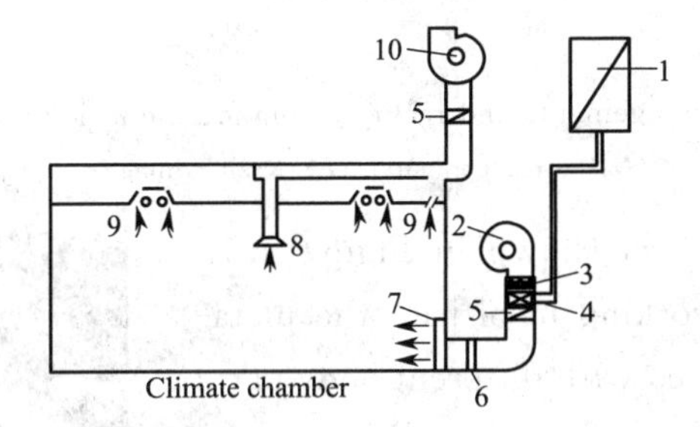

Figure 4. 5 Ventilation system of the climate chamber

1—Chiller, 2—Supply fan, 3—High efficiency filter, 4—Heat exchanger, 5—Butterfly valve, 6—Flow equalizing plate, 7—Supply air inlet, 8—Return air outlet, 9—Exhaust air outlet, 10—Exhaust fan

that part of the air is returned from a grille located at middle level and the rest of air is exhausted from a grille located ceiling level as well as openings combined with lighting fixture. In practice, it is recommended that grilles 8 and 9 are separately ducted to return and exhaust.

Two thermal manikins are arranged in the chamber to represent two working people in a small office as displayed in Figure 4. 6. Other heat sources include two computers, two lamps and two data loggers with a total heat load of 506W. The dimensions of the diffusers and the heat loads in the chamber are listed in Table 4. 1.

The dimension of the diffusers and the heat loads in the climate chamber Table 4. 1

Diffusers	Dimension(m)	Heat sources	Heat load(W)
Inlet	1×0. 6	Computer	100W×2
Return outlet	0. 5×0. 5	Manikin	80W×2
Exhaust outlet	0. 45×0. 2	Lamp	24W×4
		Data loggers	25W×2

4. 3. 2 Experimental devices and equipments

4. 3. 2. 1 Thermal manikins

The two thermal manikins were made of two male mannequins wound uniformly with

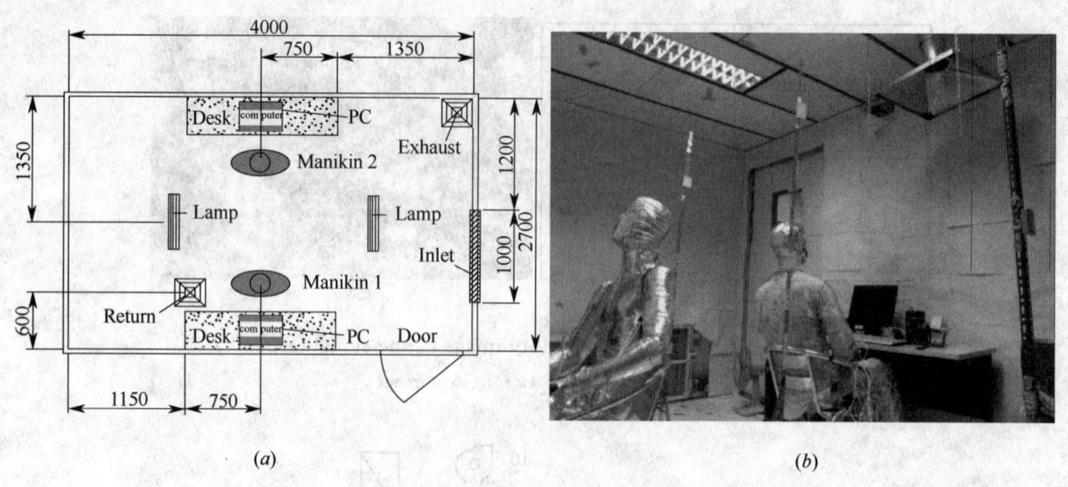

Figure 4. 6 The arrangements and reality image for the inside of climate chamber

(*a*) Arrangements; (*b*) Reality image

low-voltage electrical wires as shown in Figure 4. 7. For sedentary occupants working in office, the metabolic heat production is varied with different body parts as listed in Table 4. 2. Thus, to identify with that, the heat losses for different segments of the mannequins were controlled by using electrical wires with different resistance and length. Then, the mannequins were wrapped with tinfoil all over the body surfaces and dressed with clothing corresponds to approximately 1. 0clo as that for a typical office environment as displayed in Figure 4. 6 (*b*).

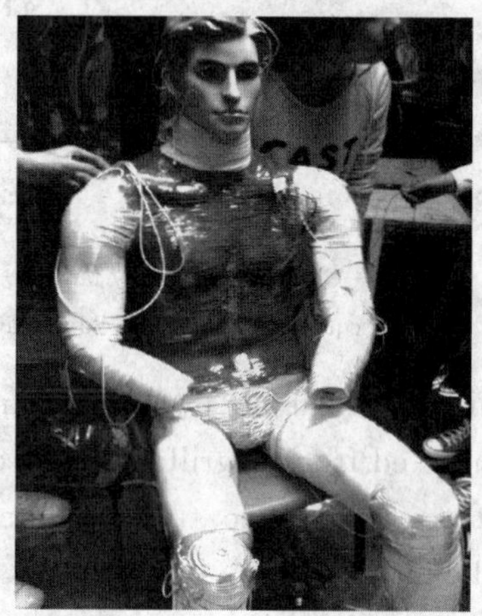

Figure 4. 7 Picture of the manikin

Heat loads for each segment of the manikin **Table 4. 2**

Segment	Heat load(W)	Segment	Heat load(W)
Head	18. 4	Right hand	0. 42
Neck	0. 43	Left thigh	4. 02
Trunk	47. 6	Right thigh	4. 02
Left fore-arm	1. 74	Left leg	2. 37
Left upper-arm	1. 18	Right leg	2. 37
Right fore-arm	1. 74	Left foot	0. 6
Right upper-arm	1. 18	Right foot	0. 6
Left hand	0. 42		

4. 3. 2. 2 Thermocouples and data loggers

Thermocouples made manually were used as temperature sensors to measure the surface temperatures of the surrounding walls and manikins. All thermocouples were calibrated first to ensure the measurement accuracy. The surface temperatures of the side walls were divided vertically into three sections to measure the surface temperatures, and the ceiling and floor temperatures were measured at four points. The inside surface temperatures of the chamber were totally measured at 38 points by using the Agilent 34970A LXI Data logger as presented in Figure 4.8 (*a*). For one of the manikins, the surface temperatures were monitored at 18 segmentsas listed in Table 4.2 with two additional monitoring points at back and one additional monitoring point at pelvis. The Graphtec America GL450 midi data logger, as shown in Figure 4.8 (*b*), with precision of ±0.1% (rdg+1.5℃) was used to digitize the body surface temperatures.

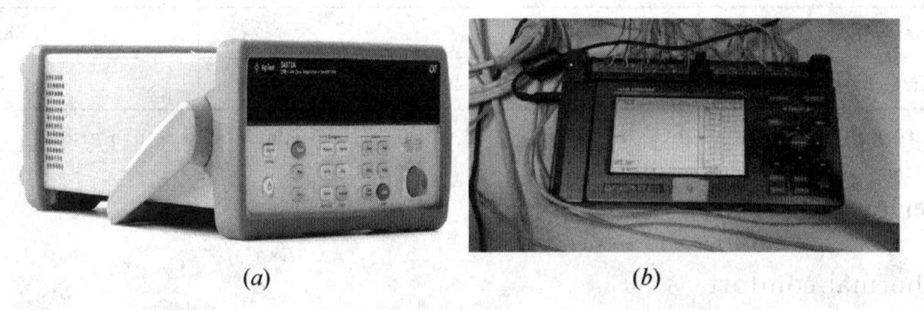

(*a*) (*b*)

Figure 4.8 Data logger for the surface temperatures of the enclosure and manikin

(*a*) 34970A LXI Data Acquisition; (*b*) Graphtec America GL450 data logger

The vertical air temperature variations were recorded at three positions (Point A, Point B and Point C) as depicted in Figure 4.9(*a*), at heights of 0.1m, 1.1m and 1.8m by using HOBO U12 Temperature Relative Humidity Data Logger. The supply, return and exhaust air temperature were all real-time monitored.

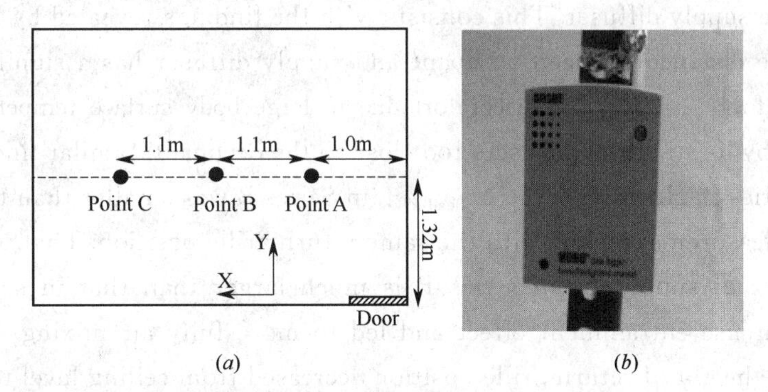

(*a*) (*b*)

Figure 4.9 Indoor air temperature measurements

(*a*) Measuring positions; (*b*) HOBO U12 data logger

4.3.3 Experimental design

All the experiments are conducted in the full-scale chamber with two manikins as displayed in Figure 4.6. The set-point temperature of the chamber is 24℃. The experimental study consists two series totally seven cases as summarized in Table 4.3. The main difference between these two series is the supply air velocity, which is much higher in Series B. The supply airflow rate in Series B is also more than that in Series A. The influences of the locations of return and exhaust grilles on the ventilation performance are investigated, in terms of thermal comfort and energy saving. What should be noted is that in Case 1 and Case 5 the return grille is combined with the exhaust grille and located at ceiling level.

Experimental study cases **Table 4. 3**

Series	A — DV				B — UFAD		
Cases	Case 1	Case 2	Case 3	Case 4	Case 5	Case 6	Case 7
Height of return inlet(m) *	2. 4	2. 05	1. 85	1. 65	2. 4	2. 05	1. 85
Supply air flow rate	84L/s				93L/s		
Return air flow rate	35L/s				45L/s		
Supply air velocity	0. 14m/s				1. 88m/s		
Supply air temperature	19℃				19. 5℃		

Height of return inlet* : The height is from floor level to the bottom edge of the return grille.

4. 3. 4 Experimental results

4. 3. 4. 1 Thermal comfort

Too large vertical temperature gradient in the occupied zone is one of the main thermal comfort issues in STARD systems [Cho et al. 2005]. In ASHRAE Standard 55 [2010], the temperature difference between head and ankle levels of occupant ($\Delta t_{\text{head-ankle}}$) is limited to 3℃. However, for the present cases in Series A, all the $\Delta t_{\text{head-ankle}}$ obtained at point A exceeded the limitation, while $\Delta t_{\text{head-ankle}}$ obtained at point B and point C were all in the accepted range, as shown in Table 4. 4. The reason is that the location of Point A is too close to the supply diffuser. This consists with the findings revealed by Lian and Wang [2002] that the distance between occupant and supply diffuser has a significant influence on thermal comfort, and thermal discomfort due to large body surface temperature difference may be causedby locating the diffusers too close to the occupant. Similar findings were also observed in Series B. However, the $\Delta t_{\text{head-ankle}}$ in Series B was smaller than that in Series A for the same measurement point with the same return grille position. That was because the momentum flux of supply air in series B is much larger than that in series A, which caused more intense entrainment effect and led to more fully air mixing in the occupied zone. When the height of return grille position decreased from ceiling level to middle level, $\Delta t_{\text{head-ankle}}$ increased slightly in both series, which meant the thermal environment was little deteriorated.

Head-to-ankle temperature differences ($\Delta t_{\text{head-ankle}}$, ℃) in different cases **Table 4. 4**

Series	A-DV				B-UFAD		
$\Delta t_{\text{head-ankle}}$ (℃)	Case 1	Case 2	Case 3	Case 4	Case 5	Case 6	Case 7
Point A	3. 4	3. 6	3. 8	3. 8	3. 1	3. 3	3. 6
Point B	2. 4	2. 7	2. 8	2. 7	2. 3	2. 3	2. 5
Point C	2. 8	2. 8	2. 7	3	1. 6	2. 3	2. 3

4.3.4.2 Energy saving and cooling coil load calculation

Table 4.5 presents the cooling coil load reductions for different heights of return grille locations in different cases, which are calculated according to Equation (4-9). Correspondingly, the cooling coil loads in different cases are also available according to Equation (4-8). In Case 1 and Case 5, the return grille was combined with the exhaust inlet at ceiling level, while in other five cases the return grille position was varied in different heights.

Energy saving for cooling coil **Table 4.5**

		Height of return grill (m)	$t^*_{1.1m\text{-}B}$	t_r	t_e	ΔQ_{coil} (W)	$\Delta Q_{coil}/Q_{Space}$
DV	Case 1	2.4	24.1	25.3	25.3	76.9	15.2%
	Case 2	2.05	24.3	24.9	25.4	82.8	16.4%
	Case 3	1.85	24.2	24.6	25.7	100.6	19.9%
	Case 4	1.65	24.3	24.3	25.8	106.5	21.0%
UFAD	Case 5	2.4	23.8	25	25	57.9	11.5%
	Case 6	2.05	23.6	24.7	25.4	81.1	16.0%
	Case 7	1.85	23.4	24.5	25.7	98.5	19.5%

$t^*_{1.1m\text{-}B}$——Air temperature at Point B with 1.1m height, ℃.

t_r——Return air temperature, ℃.

t_e——Exhaust air temperature, ℃.

The results clearly manifested that in both series, splitting the locations of return and exhaust grilles saved more energy for cooling coil. In Series A, when the height of return grille location decreased down from 2.4m to 1.65m, the return air temperature decreased and the exhaust air temperature increased. Correspondingly, energy saving of the cooling coil ΔQ_{coil} increased from 15.2% to 21% of the space cooling load. Similarly, as the air returning height decreased from 2.4m to 1.85m in Series B, the energy saving of the cooling coil ΔQ_{coil} increased from 11.5% to 19.5%.

4.4 Summary

This chapter first clearly illustrated the energy saving principles of stratified air distribution systems. The energy saving of the cooling coil in a STRAD system is associated with but differential from the occupied zone cooling load. The cooling coil load of a STRAD system is directly proportional to the exhaust air temperature. In a STRAD system, splitting the locations of return and exhaust grilles may have slightly effects on the occupied zone cooling load, but strongly increases the energy saving of the cooling coil. A novel cooling coil load calculation method for STRAD systems was developed based on an available exhaust air temperature. For the first time, experiments were also conducted in a STRAD system with separate locations of return and exhaust grilles. The experimental results indicated that the temperature difference between the head and ankle levels $\Delta t_{head\text{-}ankle}$ increased slightly when the return grille was lowered from ceiling level to middle level. The experimental data also demonstrated that splitting the locations of return and exhaust grilles, further energy saving of the cooling coil was attained. As the height of return grille location decreased from 2. 4m to 1. 85m, the return air temperature decreased and the exhaust air temperature increased in both series of experiments. Correspondingly, energy saving of the cooling coil ΔQ_{coil} increased from 15. 2% to 19. 9% of the space cooling load in Series A and from 11. 5% to 19. 5% in Series B. The experimental results confirmed that the newly developed cooling coil load calculation method is feasible to be used in STRAD systems and CFD simulations are required at the design stage to obtain the exhaust air temperature.

Chapter 5 Stratified air distribution in a small office

5.1 Introduction

There are seven types of models that able to predict the ventilation performance in buildings, which are differential in details among different models [Chen 2009]. The full-scale model is the most reliable one but expensive and time consuming, while the CFD model is the most sophisticated one but capable of conveniently providing the most detailed information about the flow fields [Chen et al. 2010]. As reviewed in Section 2.6, at present all the CFD models use approximations to predict the ventilation performance in buildings. The approximations inevitably bring some uncertainties in the predicted results, for instance, the distributions of airflow fields and chemical-species concentrations. In addition, the applications of CFD models are selective and there is no one that suitable for all flow conditions with different features. Thus, before using a CFD model to assess the performance of a specific ventilation system applied in buildings, a validation procedure is required by using experimental data or reliable semi-empirical correlations [Chen and Srebric 2002]. Therefore, in this Chapter, a CFD model is validated first by using the experimental results presented in Chapter 4 as well as the data from published literature. Then the CFD model is going to be adopted in the following numerical studies.

In Chapter4, the influences of the diffuser's locations to the performance of STRAD systems have been experimentally investigated from viewpoint of thermal comfort and energy saving. However, due to the limitations of experimental conditions, the height of return grille position can only be adjusted in the upper zone within a limited range. In addition, the effective cooling load factors for each heat source with different return heights are difficult to be obtained in the experiments. Furthermore, the indoor airflow in an office room located at a perimeter zone can be greatly affected by the outdoor climate. Thus, the influences of diffuser's locations to the performance of a STRAD system in perimeter zone may be differ widely from that in the climate chamber and should be further studied. Therefore, a hypothetical small office located at a perimeter zone ventilated by a STRAD system with separate locations of return and exhaust grille is studied by using the validated CFD model. The height of return grille position is varied from ceiling level to floor level. The effects of the variation on the thermal environments and cooling coil load reduction are investigated. The influence of outdoor climate to the indoor air flow characteristic is also considered. Furthermore, the cooling load factors for each heat sources located in the office environment are calculated based on simulation results.

5.2 CFD model validation

By using the experimental results presented in Chapter 4, a CFD model validation is conducted. Correct simulation of airflow in a room depends on proper specifications of boundary conditions and selection of turbulence model. The standard $k-\varepsilon$ model can simulate convective heat transfer of buoyancy-driven air flow, as long as a reasonable value of y^+ is achieved. Hence, it is adopted in the simulations, together with a differential viscosity model to account for low Reynolds number effect. The numerical schemes and boundary conditions are summarized in Table 5.1.

The details of the numerical methods **Table 5.1**

Turbulence Model	Standard $k-\varepsilon$ model
Numerical Schemes	Staggered third order PRESTO scheme for Pressure; Upwind second order difference for other terms; SIMPLE algorithm
Floor, ceiling, side walls	Fixed surface temperatures according to the measured dates
Computer	Uniform heat flux; 100W
Tables	Adiabatic wall
Lightings	Uniform heat flux; 24W×2
Data loggers	Uniform heat flux; 25W×2
Human Body	Fixed skin temperatures of 18 body segments according to the experimental results
Supply air outlet	Velocity inlet (84L/s, 19℃)
Return air inlet	Velocity inlet with a negative direction (35L/s)
Exhaust air inlet	Pressure-outlet
Radiation heat	Discrete Ordinates (Do) radiation model

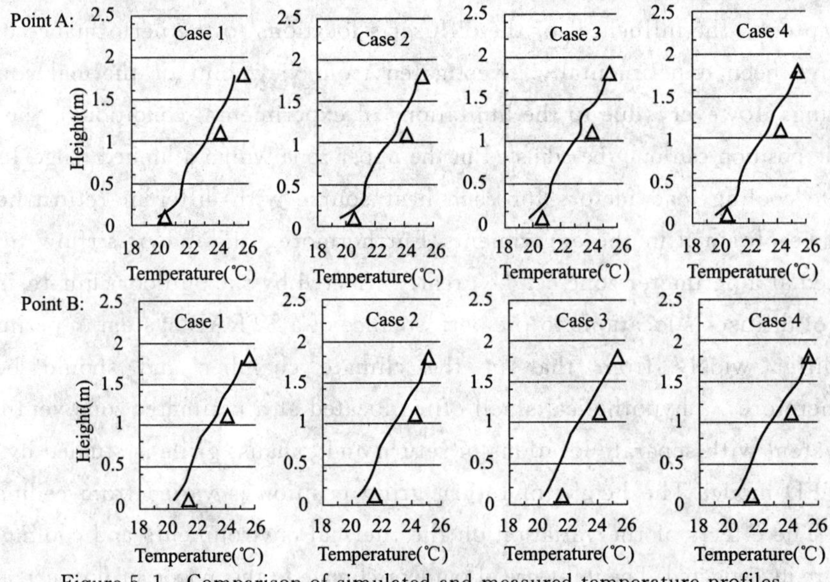

Figure 5.1 Comparison of simulated and measured temperature profiles

Triangle symbols—measured temperature; solid lines—simulated temperature

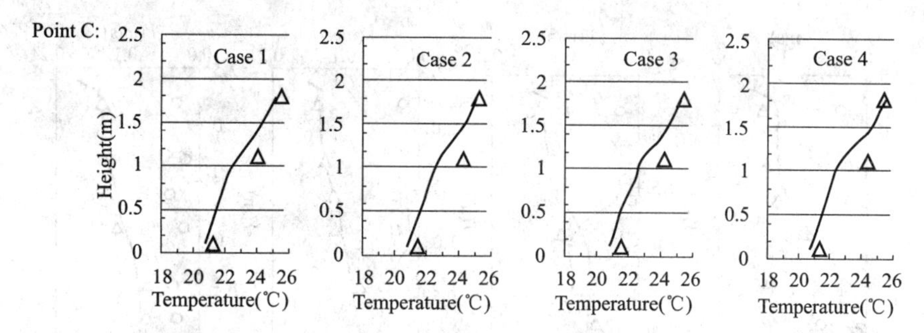

Figure 5.1 Comparison of simulated and measured temperature profiles (continue)

Triangle symbols—measured temperature; solid lines—simulated temperature

Figure 5.1compares the measured and simulated air temperature profiles at different positions in the test chamber. At middle level (1.1m) of the room, the simulated air temperature is slightly lower than the measured. It is mainly because that the height of the thermal manikin adopted in simulations is somewhat higher than that used in experiments. As indicated in REHVA guidebook No.1 [2002], elevating heat source from lower part of the room to the upper part, the temperature gradient would be reduced in lower part and increased in the upper part. The measured and simulated data agreed well for most conditions.

5.3 Further validation for the CFD model

An experiment carried out by Zhang and Chen [2006] in a full-scale environmental chamber with under floor air distribution (UFAD) system was selected to further validate the computational model.

Temperature is normalized as $T=(t-t_s)/(t_e-t_s)$. V1 to V7 represent seven horizontal positions in the original experiments [Zhang and Chen 2006].

Figure 5.2 displays the comparison results of vertical temperature and velocity profiles between the numerical simulation and experiment. For most points, the velocity and temperature distributions were in good agreements. Thus, it is believed that the selected CFD model is capable to evaluate the ventilation performance of STRAD systems with similar layout.

5.4 Simulation cases

The hypothetical small office is located at a perimeter zone with a dimension of 4.0m (L)×3.0m(W)×2.7m(H), as shown in Figure 5.3. Heat objects in the office include two occupants, two computers, two lamps and a printer. The cooling loads for each heat source including the exterior wall are summarized in Table 5.2. For the exterior wall, the window to wall ratio is 20% and the cooling load is fixed at 39W/m^2 according to the guide of overall thermal transfer value (OTTV) for Hong Kong [1995]. The required fresh air

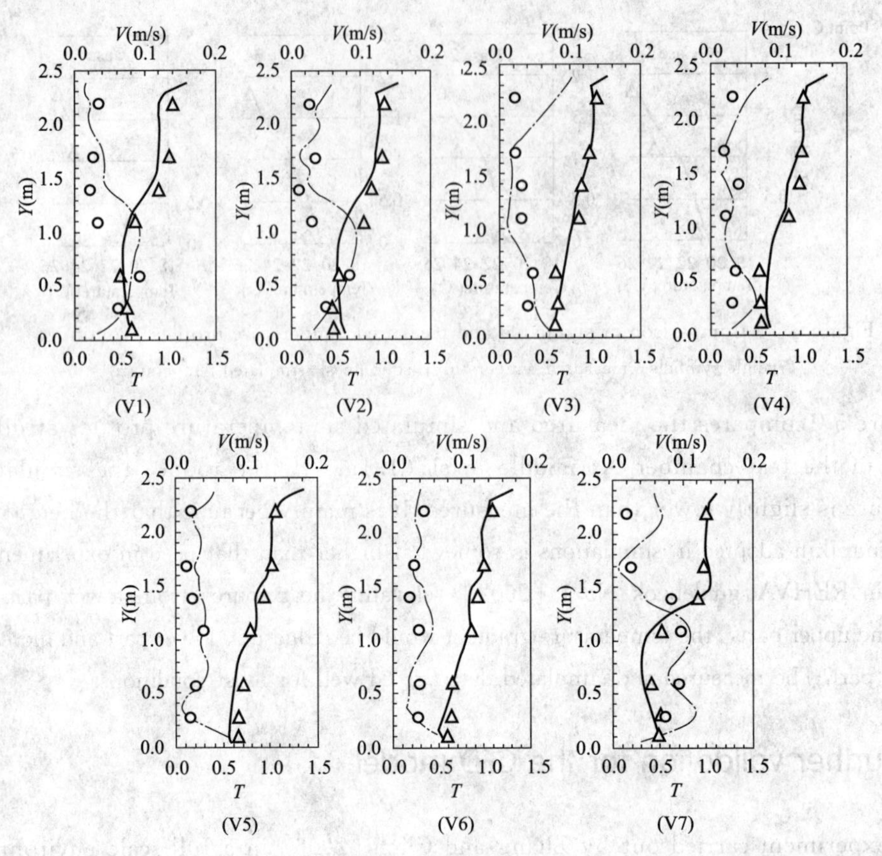

Figure 5.2 Comparison of simulated and measured temperature and velocity profiles
rectangular symbols—measured temperature after normalization; triangle symbols—measured velocity;
solid lines—simulated temperature; dash line—simulated velocity

for an occupant is 10L/s. The air is supplied from two swirl diffusers located at floor level with a flow rate of 0.12m^3/s and a temperature of 18℃. The set-point air temperature for the room in the occupied zone is 25℃. 16.7% of the total air is directly exhausted from a grille located at ceiling level and the rest air is returned from side wall. The location of return grille is varied from ceiling level to floor level in different cases as listed in Figure 5.4 and correspondingly seven cases are simulated as shown in Table 5.3.

Cooling loads for the office **Table 5.2**

Heat Sources	Occupants	Computers	Lightings	Window	Exterior wall	Printer	Total
Cooling Load	80W×2	100W×2	70W×2	84W	336W	100W	1020W

Different simulation cases **Table 5.3**

Simulation Cases	Case 1	Case 2	Case 3	Case 4	Case 5	Case 6	Case 7
Height of return inlet* (m)	2.6	2.3	2.0	1.7	1.3	0.8	0.3

Height of return inlet*: The height is from floor level to the bottom edge of the return grille.

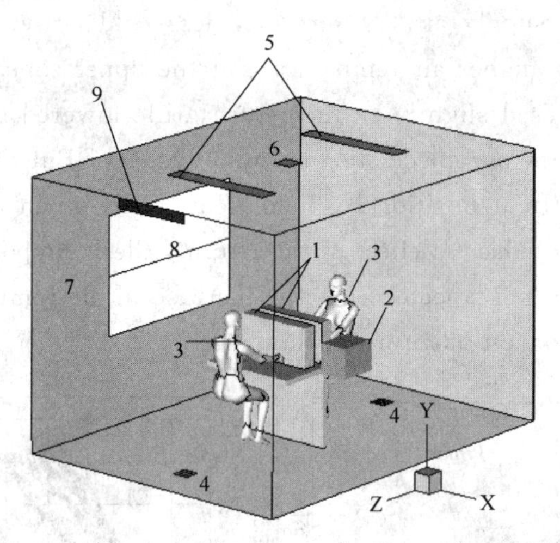

Figure 5. 3 The simulation model——a small office with exterior wall (Case 1)

1 — Computer; 2 — Heat source; 3 — Occupants; 4 — Diffusers; 5 — Lighting;

6 — Exhaust outlet; 7 — Exterior wall; 8 — Window; 9 — Return outlet

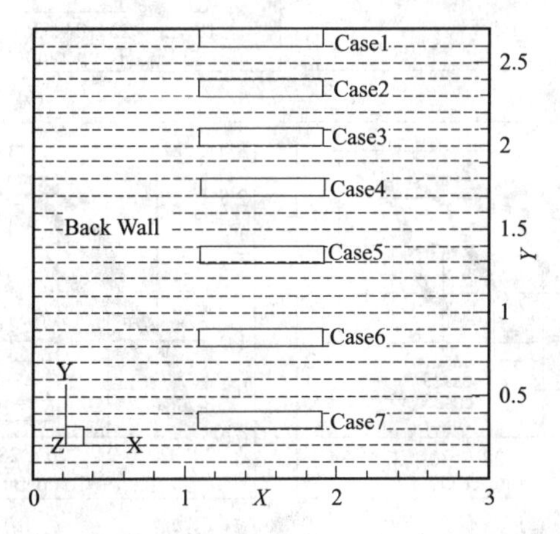

Figure 5. 4 Heights of return grille in different cases

5. 5 Simulation Results

5. 5. 1 Thermal comfort

As shown in Figure 5. 5, the temperature distributions from floor level to ceiling level are reported at three positions: points A and B of 0. 2m from the side of occupants, and point C that 0. 2m away from the desk. It can be observed that as the height of return grille position decreased, the room air temperature increased in the upper zone of the room, which was especially obvious in Case 6 and Case 7. That was because part of the cold supply air was returned directly before mixed with the indoor air as revealed in Figure 5. 6 at region 2. The lower the height of return grille position, the more significant the "short-circuit" of cold supply air. As a result, less cold air was entrained by thermal plumes, which led to a higher air temperature at the upper zone. The air temperature in the occupied zone increased slightly as the return grille lowered from the ceiling level to the upper boundary of the occupied zone (1. 3m in Case 5) . But the air temperature in the occupied zone, especially at Position B, increased rapidly when further decreased return grille location. Since the induction effect of the return grille is proportional to the momentum flux of return air, it can be expected that the increase of air temperature in the occupied zone is related to the portion of return airflow rate.

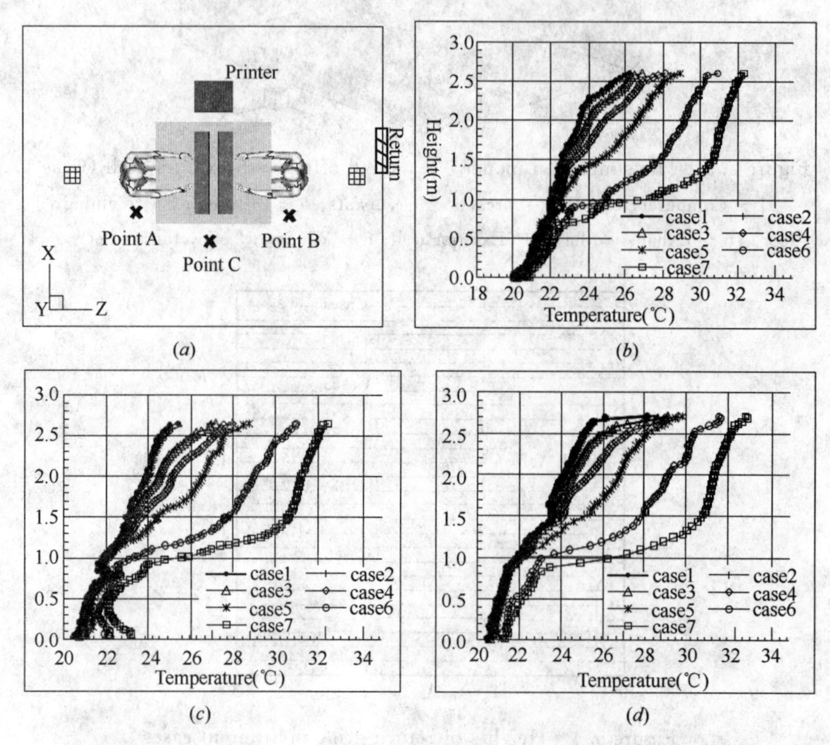

Figure 5. 5 Temperature stratifications in different cases

(*a*) Temperature monitored positions; (*b*) Temperature stratification — Point A;

(*c*) Temperature stratification — Point B; (*d*) Temperature stratification — Point C

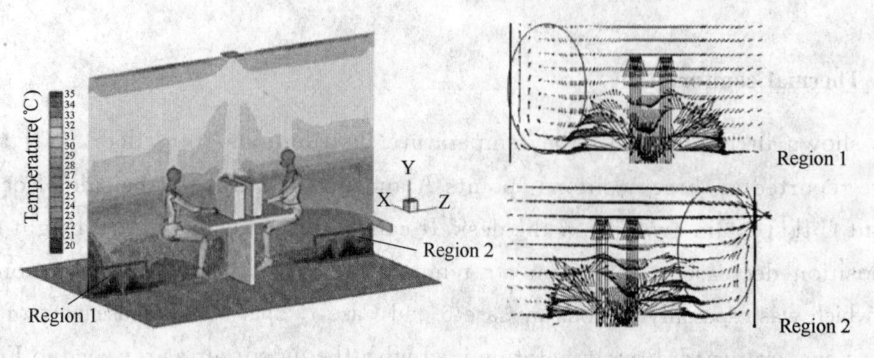

Figure 5. 6 Temperature and velocity distribution at Plane X=1. 5m in Case 7

The temperature differences between head and foot levels ($\Delta t_{head\text{-}foot}$) at the three positions are shown in Table 5.4. It can be found that when the return grille was located above the upper boundary of the occupied zone (1.3m in Case 5), the values of $\Delta t_{head\text{-}foot}$ were all in the acceptable range. This is consistent with the conclusions obtained from previous experiments. Further lowering the return inlets into the occupied zone, the room air temperature at the head level increased much more than that at foot level, which may adversely affect the thermal comfort. For instance, when the air was returned at 0.3m above the floor as in Case 7, the $\Delta t_{head\text{-}foot}$ at position A was 6.9 ℃ and far beyond the 3℃ limit.

Head-to-ankle temperature differences in different cases **Table 5.4**

$\Delta t_{head\text{-}ankle}$	Case 1	Case 2	Case 3	Case 4	Case 5	Case 6	Case 7
Point A	1.7	1.8	1.8	1.8	2.1	4.4	6.9
Point B	1.5	1.5	1.5	1.6	1.6	2.0	4.0
Point C	1.5	1.5	1.6	1.7	1.7	3.1	6.3

5.5.2 Energy saving

According to space cooling load calculation method presented in ASHRAE Standard [2011], the convection part of heat gains become to be cooling load immediately. While the radiation heat is absorbed by furniture and interior room surfaces first, and then converted into cooling load later only when it is transfer to indoor air by convection from these surfaces. Therefore, different with the method described in Equation (2-8), the occupied zone cooling load $Q_{i-occupied}$ also can be calculated as the following equation:

$$Q_{i-occupied} = Q_{convective,i-occupied} + Q_{conv,surface-occupied} \tag{5-1}$$

where $Q_{convective,i-occupied}$ is the convection heat transfer directly to the indoor air from heat source i located in the occupied zone; $Q_{conv,surface-occupied}$ is the radiation heat absorbed by furniture and interior room surfaces that located in the occupied zone and then transferred to the indoor air by convection. The equation indicated that the distribution percentage of heat gains in convection and radiation portions directly influenced the occupied zone cooling load.

Table 5.5 presents the radiation heat portions for each heat source in different cases. The increase of radiation heat portion for the exterior wall indicated the convection heat portion decreased, which might benefit to reduce the occupied zone cooling load. While for the heat sources located in the occupied zone, including the heat source, the computer and the occupants, the radiation heat portions decreased, which implied that the convection heat transfer between these heat sources and the indoor air were enhanced.

Radiation heat portions for each heat source in different cases **Table 5. 5**

	Case 1	Case 2	Case 3	Case 4	Case 5	Case 6	Case 7
Exterior wall	59. 8%	60. 4%	60. 7%	61. 6%	63. 2%	64. 9%	67. 5%
Windows	63. 1%	63. 1%	63. 3%	64. 5%	67. 9%	77. 4%	83. 7%
Heat source	72. 8%	72. 7%	72. 5%	72. 0%	71. 7%	70. 0%	68. 0%
Computer	49. 4%	47. 0%	46. 3%	45. 9%	45. 7%	45. 5%	45. 0%
Occupants	32. 0%	31. 0%	29. 5%	27. 6%	25. 4%	22. 6%	20. 6%
Lamps	81. 5%	81. 5%	82. 8%	83. 1%	83. 1%	83. 4%	84. 2%

With separate locations of return and exhaust grilles, it is possible to further reduce the cooling coil load. As revealed in Section 4. 2, in a STRAD system the energy saving for cooling coil was associated with but differential from the space cooling load reduction. Table 5. 6 illustrates the influence of the height of return grille position on the reduction of cooling coil load. In Case 1, when the return grille was located at ceiling level as the common layout in practical applications, the energy efficiency of the ventilation system was the worst. The cooling coil load in Case 1 was even slightly greater than that in a MV system, since part of thermal plume with relatively high temperature directly entered the return grille and contributed to the coil load, which can be clearly observed in Figure 5. 7.

Energy saving for cooling coil **Table 5. 6**

	Height of return grill (m)	Return Air Temperature (℃)	Exhaust Air Temperature (℃)	ΔQ_{coil} (W)	$\Delta Q_{\text{coil}}/Q_{\text{Space}}$
Case 1	2. 6	25. 1	24. 8	−4. 8	−0. 5%
Case 2	2. 3	24. 9	25. 1	2. 4	0. 2%
Case 3	2. 0	24. 8	26. 1	26. 6	2. 6%
Case 4	1. 7	24. 6	27. 0	48. 3	4. 7%
Case 5	1. 3	24. 2	28. 8	91. 7	8. 9%
Case 6	0. 8	23. 6	30. 9	142. 4	14. 0%
Case 7	0. 3	22. 5	32. 3	176. 3	17. 3%

As the return grille was lowered, the exhaust air temperature increased and much more convective heat was immediately drawn into the exhaust grille. As shown in Figure 5. 8 for Case 5, the free convective flow along exterior wall brings the released heat to the upper zone. Thus, the energy saving of the cooling coil was significantly increased to 17. 3% of the space cooling load was saved in Case 7. The results were in line with the experimental conclusions of Chapter 4. However, the relative energy saving of cooling coil ($\Delta Q_{\text{coil}}/Q_{\text{Space}}$) in numerical studies was not as significant as that in experimental conditions. That was because the simulated office was in the perimeter zone, its space cooling load was larger than that in the climate chamber. In addition, it should be noted that in simulation cases there were no openings for exhaust air on the lighting fixture and that the air was

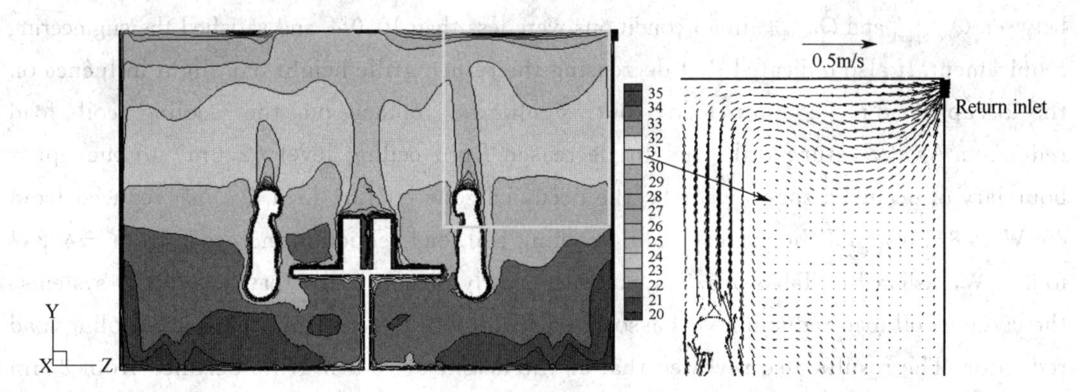

Figure 5.7 The temperature and velocity distribution at Plane $X=1.5$m in Case 1.

exhausted only from grille located at ceiling level, which was different from experiments as shown in Figure 4.2.

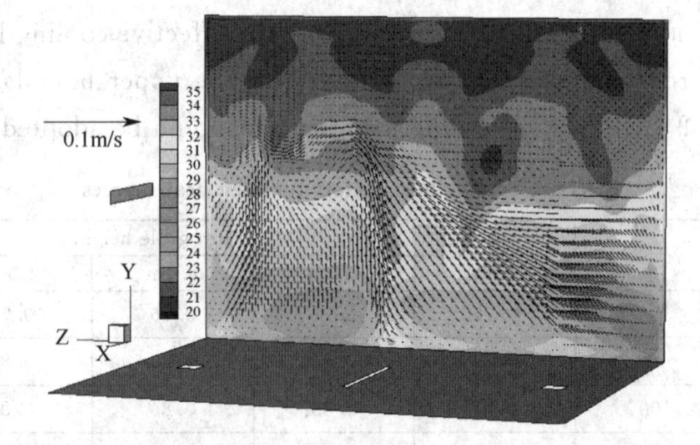

Figure 5.8 The temperature and velocity distribution at Plane $X=0.01$m in Case 5

5.5.3 Effective cooling load factors

To obtain the effective cooling load factors for each heat source with different return grille heights, additional cases were simulated. In Section 5.5.1 the thermal environments have been demonstrated to be poor when the return grille was located in the occupied zone. Therefore, in this section only the conditions that the return grille located above the occupied zone were simulated. In other words, there totally 30 additional cases for the six heat sources with five return grille heights, varied from 2.6m in Case 1 to 1.3m in Case 5, were simulated and the effective cooling load factors were calculated. The occupied zone cooling loads Q_{occupied} were calculated according to Equation (2-8), while the effective cooling loads factors were calculated according to Equation (2-7). Based on that, the effective occupied zone cooling loads were able to be obtained according to Equation (2-12). Table 5.7 presents the effective cooling load factors for each heat sources. It revealed that the effectiveness of the cooling load calculation method proposed by Xu et al. (2009) was verified for the simulation conditions with different return grille heights. The divergences between $Q'_{occupied}$ and $Q_{occupied}$ in all conditions were less than 10. 0% and satisfied the engineering requirement. It also indicated that decreasing the return grille height has slight influence on the occupied zone cooling load, but significant impact on the cooling coil load reduction. As the return grille height decreased from ceiling level (2. 6m) to the upper boundary of occupied zone (1. 3m), the occupied zone cooling load $Q_{occupied}$ reduced from 856W to 841W, and the corresponding cooling coil load reduction increased from −4. 8W to 91. 7W, aslisted in Table 5. 5. The statistics clearly demonstrated that in STRAD systems, the cooling coil load reduction was associated with but different from the space cooling load reduction. The results also revealed that as the height of return grille declined from 2. 6m to 1. 3m, the effective cooling load factors for all the six heat sources decreased slightly. Therefore, the effective cooling load factors for each heat sources, which are summed in Table 5. 7, can be adopted directly according to the height of return grille, when designing a STRAD system with similar configuration in an office environment. When the return grilles are designed at ceiling level, high values of the effective cooling load factors should be used. When the return grilles are located close to the upper boundary of the occupied zone, low values of the effective cooling load factors should be adopted.

Effective cooling load factors for each heat sources **Table 5. 7**

$ECLF_i$	Q_i (W)	Return grille height				
		2. 6m	2. 3m	2. 0m	1. 7m	1. 3m
Exterior wall	336	0. 77	0. 77	0. 76	0. 75	0. 74
Windows	84	0. 83	0. 80	0. 80	0. 81	0. 80
Heat source	100	0. 87	0. 85	0. 84	0. 81	0. 81
Computer	200	0. 90	0. 91	0. 90	0. 88	0. 89
Occupants	260	0. 96	0. 95	0. 93	0. 92	0. 89
Lamps	140	0. 67	0. 66	0. 65	0. 65	0. 64
$Q'_{occupied}$ (W)		940	931	920	908	897
$Q_{occupied}$ (W)		856	855	852	844	841
$(Q'_{occupied}-Q_{occupied})/Q_{occupied}$		9. 8%	8. 9%	8. 0%	7. 6%	6. 7%

5. 6 Summary

In this Chapter, a CFD model validation procedure was carried out first by using the experimental results presented in Chapter 4. In monitored cases, the measured and simulated data agreed well, which indicated that the accuracy of the simulated results was acceptable for indoor environment design. Furthermore validation for the CFD model was also conducted by using experimental results published in a previous literature. The comparison results of the velocity and temperature distributions between the experiment and the simulation confirmed that this CFD model was capable to evaluate the ventilation performance of

STRAD systems.

Based on that, the application of a stratified air distribution (STRAD) system in a small office with low ceiling level (3m) was numerically studied. The simulation results manifested that the newly developed CFD-based cooling coil load calculation method in Chapter 4 is valid to evaluate the energy saving of the cooling coil and determine the cooling coil load when designing a SATRD system. The effects of return and exhaust grilles locations on the performances of the STRAD system have been investigated. Extra energy saving of the cooling coil is achieved by splitting the exhaust and return grilles and installing the return grille at middle level. Maintaining the exhaust grille at ceiling leveland decreasing the return grille location from ceiling level to lower zone, the occupied zone cooling load reduced slightly, but the energy saving ofthe cooling coil increased obviously. The coiling coil load reduction was directly proportional to the exhaust air temperature. The lower the return grille, the higher the exhaust air temperatureand the more the energy saved for the cooling coil. As much as 17.0%of the space cooling load was saved for the cooling coil when the return grille was located at 0.3m above floor level. The thermal environment in the occupied zone was slightly affected as the height of return grille location decreased from ceiling level to the upper boundary of the occupied zone (1.3m for sedentary occupants). However, further lowering the return grille into the occupied zone and locating it too close to the supply diffusers may lead to "short-circuit" of cold supply air and have a negative effect on the thermal environment. Therefore, considering both thermal comfort and energy saving, it is recommended to locate the return grille at the upper boundary of the occupied zone. The exhaust grille should be located at ceiling level and close to the lighting fixture and exterior wall, which may facilitate the direct exhaust of some convective heat. The effective cooling load factors $ECLF_i$ for each heat sources with different heights of return grille locations were provided in Table 5.6 based on 30 additional simulation cases, which could be adopted conveniently by the consultant engineers to calculate the occupied zone cooling load and then determine the required air flow rate.

Chapter 6 Stratified air distribution system in a large lecture theatre

6.1 Introduction

Several researches [Karimipanah et al. 2000, Fong et al. 2011] have been investigated the applications of STRAD systems in classrooms, and demonstrated that the STRAD system was able to improve the thermal comfort and ventilation effectiveness, and thus beneficial for the learning potential of students and also their health [Wargocki et al. 2005, Shaughnessy et al. 2006]. However, their studies were limited to general classroom with flat floor and common ceiling level (no more than 3m). The applications of STRAD systems in large space lecture theatres with terraced floor still rarely reported. In a lecture theatre with large space and terraced floor, portion of cold supply air from floor level with low-velocity may tends to flow downwards along with the terraced floor, leading to undesirable temperature stratifications from front to back rows of seating, and causing overcooling in the front row and inadequate cooling in the back row. Furthermore, the internal heat gains in a large-space lecture theatre are much more than that in a general classroom, which result in more complex indoor air flow. Thus, the thermal environment in a large-space lecture theatre with terraced floor that ventilated by a STRAD system is much more complicated than that in a general classroom with flat floor. The findings for the latter may not suitable for the former and cannot be adopted directly.

In this chapter, to achieve satisfied thermal environment for the students, numerical investigations are conducted on the application of STRAD system in a terraced lecture theatre with large space. Since the numerical thermal manikin is useful to evaluate local thermal environment in CFD simulations, it is adopted in present study. However, the amount of numerical thermal manikin used in a CFD simulation is limited to the computational capacity of computer, since mesh number generated for a numerical thermal manikin is huge and time-consuming. Therefore, only one numerical thermal manikin is used in the simulation cases. Three cases are simulated first with correspondingly three different locations of the numerical thermal manikin. The local thermal comfort in different locations are assessed and compared by using the UCB thermal comfort model [Zhang 2003], in order to find the region with worst local thermal environment. Based on that, different designs to realize thermal stratified distribution in the lecture theatre are studied to optimize the thermal environment and maximize the energy saving potentials. The cooling load calculation method proposed by Xu et al. [2009] is tested as well for the simulation conditions. The cooling coil reductions generated due to vertically separating exhaust and return air grilles are also evaluated by using the cooling coil load calculation method developed in Chapter 4. Then, the influence of different return locations on the performance of the STRAD system, when applied in the lecture theatre, is investigated as well. The effective cooling load factors for each heat source in different conditions are calculated and compared.

6.2 Lecture theatre configurations

As shown in Figure 6.1(*a*), a large lecture theatre in terms of dimension, occupant density and lighting layout referred to an existing classroom in the university campus is selected. There are 130 seats arranged in 10 rows on a terraced floor in the classroom. The height from ceiling to floor of the classroom varies from 5m at first row to 3m at tenth row. 20 lamps are installed at the ceiling level and simplified as rectangle panels. The existing air-conditioning system is a fan-coil unit (FCU) + primary air-handing unit (PAU) design, with ceiling supply and ceiling returns. In the numerical study of this chapter, several hypothetical STRAD air distributions are simulated. As indicated in Figure 6.1(*b*), the air is supplied into occupied zone directly, i.e. from floor-level, terrace or desk edge. Part of the air is exhausted from three grilles located at ceiling-level. What notable is that different from conventional design, the rest four return grilles are distributed on surrounding walls at middle-level, as indicated in Figure 6.1(*a*).

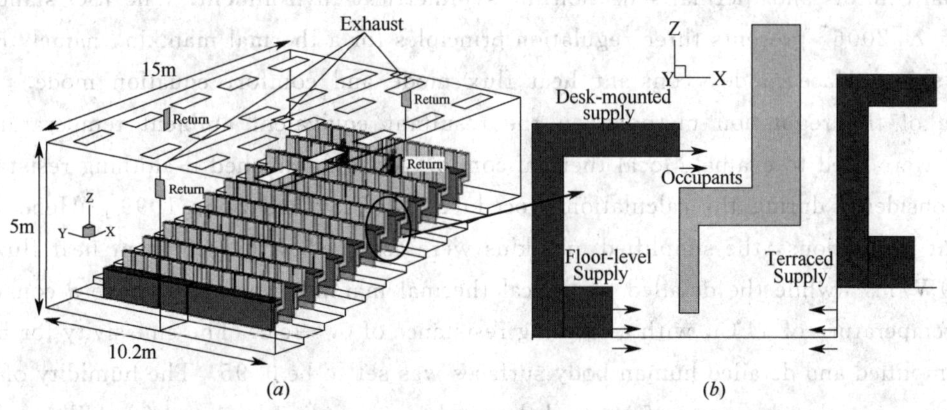

Figure 6.1 Schematic diagram of the classroom model in the simulation cases and side view of different air supply locations

(*a*) Schematic diagram of the classroom model; (*b*) Different air supply locations

6.3 Mesh generation and Boundary conditions

Due to the symmetrical characteristic of the lecture theatre, half of the room was selected as the simulation domain in order to save the computational time. Computational domain was divided into several sub-regions and most regions were meshed by structured grid with high quality. The total mesh number was about 2.3 millions. Since the lecture theatre was located at interior zone, it was assumed that lamps and occupants were the only heat sources in the space. According to the ASHRAE Standard [2009], the lighting power density was set to be 15W per ceiling area. The instantaneous heat gain from lighting can be calculated as following:

$$Q_{el}=WF_{ul}F_{sa} \tag{6-1}$$

where, Q_{el} (W) is the heat gain, W (W) is the total light wattage, F_{ul} is the lighting use factor and F_{sa} is the lighting special allowance factor. The depth of the lecture theatre is 15m and some lamps are required to be turned off during class time. Hence, the lighting use factor was set to 0.75 and the lighting special allowance factor was set to 0.9 for a common fluorescent lamp. In reality, different types of lamp have varied luminous efficiency as well as the fraction of visible lighting and heat [Thimijan and Heins 1983]. The lamps were simplified as panels with fixed heat flux and the total instantaneous heat gain of lightings was 764W.

In order to save the calculation time, all occupants except one are represented with simplified geometry as shown in Figure 6.1(*b*). The surface area of each simplified human body is 1.78m^2. The influence of a human body on the room air flow and air temperature is able to be simulated by such a simplified geometry [Topp et al. 2002]. However, a thermal manikin with realistic body geometry is required to evaluate local thermal comfort and thermal sensation in asymmetrical environment. The ISO standard 14505-2 [2006] presents three regulation principles for a thermal manikin, namely constant skin surface mode, constant heat flux mode and comfort equation mode. The impact of the regulation methods on the resultant equivalent ambient temperatures, which were used to evaluate local thermal comfort, was diminished if clothing resistance was considered during the calculation procedure [Melikov and Zhou 1999]. Hence, in present simulations, the simplified manikins were assigned with a constant heat flux of 39.19W/m^2, while the detailed numerical thermal manikin was set to have a constant skin temperature of 34℃, with a clothing resistance of 0.59clo. The emissivity for both the simplified and detailed human body surfaces was set to be 0.95. The humidity of the air was assumed to be kept at 50% and the total space cooling load was 69.4W per floor area.

The zone outdoor airflow V_{oz}, which defined as the outdoor airflow that must be provided to the zone by the air distribution system, was calculated according to ASHRAE Standard 62.1 [2010]:

$$V_{oz}=V_{bz}/E_z \tag{6-2}$$

where, V_{bz} is the required outdoor airflow in the breathing zone of the occupiable space; E_z is the zone air distribution effectiveness; V_{bz} is determined by the outdoor airflow rate required per person R_p and the outdoor airflow rate required per unit area R_a. R_p was set to 6L/s • person and R_a was set to 0.6L/s • m^2 according to ASHRAE Standard 62.1 [2010]. The E_z was set to 1.2 for thermal stratification air distribution system. Hence, the zone outdoor airflow V_{oz} was 359L/s. As input for the CFD simulation, the total supply air volume was calculated to be 565.4L/s and the supply air temperature was constant at 18℃, based on an experienced presumption that the occupied zone cooling load accounted for 90% of the whole space cooling load.

6.4 Local thermal environments in different rows

To compare local thermal environments in different rows, three cases were simulated first. A numerical thermal manikin was respectively located at first, fifth and ninth row of the space in these three cases. All of the air was supplied from the floor-level diffusers as shown in Figure 6.1(*b*), with a constant velocity of 0.3m/s. Figure 6.2 presents the local thermal environment evaluation results in different cases. As expected, for head level the thermal environment at back row was much closer to the warm side than that at front row. While for the foot level, occupants seating at back rows was suffered from local cold draft. The overall thermal comfort for the occupants seated at back rows was worst. Thus, it was considered that in the terraced lecture-theater ventilated by a STRAD system, thermal environment at back rows was much poorer than that at front rows. Therefore, in subsequent studies, the numerical thermal manikin was located at the ninth row of the room as shown in Figure 6.1(*a*), and the thermal environments at back rows were concerned and optimized.

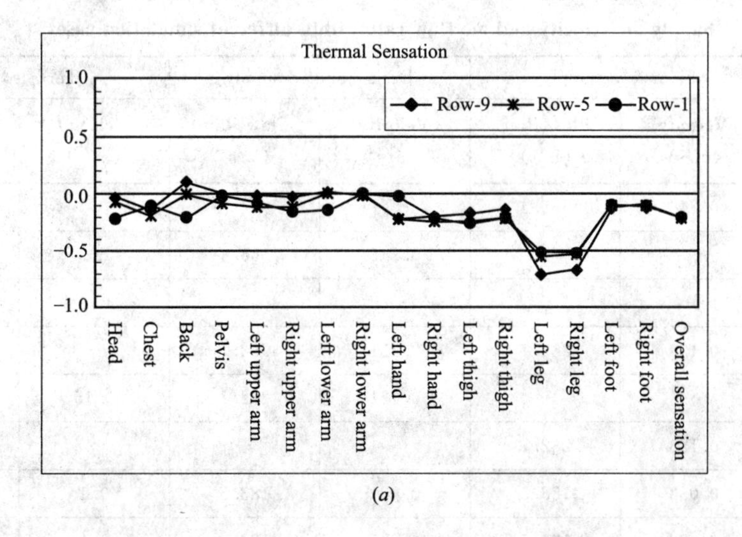

(*a*)

Figure 6.2 Thermal sensation and comfort for the numerical thermal manikin located at different rows

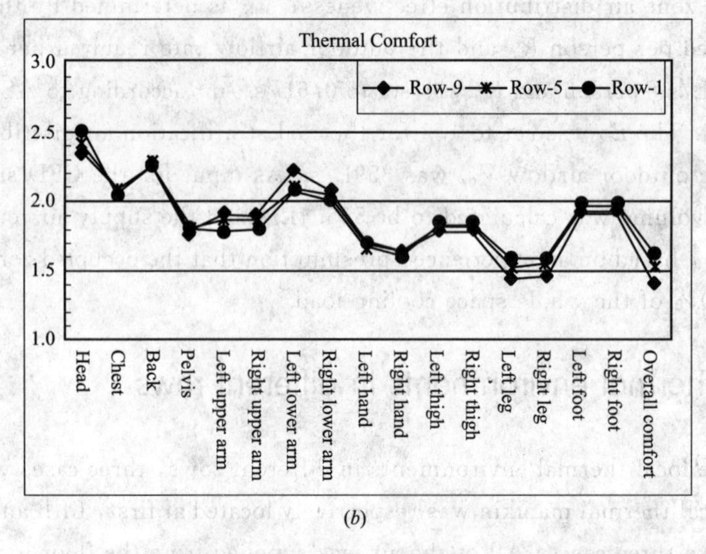

(*b*)

Figure 6.2 Thermal sensation and comfort for the numerical thermal manikin located at different rows (continue)

6.5 Different stratified air distribution designs

6.5.1 Simulation cases description

Different designs to realize thermal stratified distribution in the space were simulated. The combinations of the supply air locations, as presented in Figure 6.1(*b*), are summarized in Table 6.1.

Supply air velocity and air flow rate within different simulation cases **Table 6.1**

Supply air locations	Floor-level supply		Desk-mounted supply		Terraced supply	
	Velocity (m/s)	Flow Rate (L/s)	Velocity (m/s)	Flow Rate (L/s)	Velocity (m/s)	Flow Rate (L/s)
Case 1	0.246	565.4	—	—	—	—
Case 2	—	—	0.242	565.4	—	—
Case 3	—	—	—	—	0.3	565.4
Case 4	0.123	282.7	0.121	282.7	—	—
Case 5	—	—	0.121	282.7	0.15	282.7
Case 6	0.123	282.7	—	—	0.15	282.7
Case 7	0.082	188.5	0.081	188.5	0.1	188.5
Case 8	0.082	188.5	0.081	188.5	0.1	188.5
Case 9	0.082	188.5	0.081	188.5	0.1	188.5

To avoid potential drafts, the supply air velocities were relatively low. The first seven cases were simulated to optimize the thermal environment in the terraced lecture-theater. Based on that, two additional cases were conducted to obtain the respective effective cooling load factors for occupants load and lighting load. For the first seven cases, the cooling load was comprised of heat released from occupants and lightings. While for the two additional cases, there was only one type of heat source, either occupants or lightings. For instance, in Case 8 the occupants were the only heat source.

6.5.2 Airflow and temperature stratification

Figure 6.3 presents the velocity vectors and temperature profiles at plane $y = 4\text{m}$, which is the section that pass through the detailed thermal manikin. Compared the flow fields in different cases, one main notable common feature was the obvious temperature stratifications in all cases. What interesting was that in these cases, significant temperature gradients also existed in horizontal direction in the occupied zone. This was attributed to the thermal air plumes flowed from front-row to the back rows along the head height of occupants. Thus, among different rows, the upper body parts of the occupants seated at back rows were exposed to relatively high temperature air, while the air temperature at floor level was almost the same. Therefore, it was much likely for occupants seated at back rows to have the "warm-head and cold-feet" feelings, which was an undesirable thermal comfort condition. This is consists with the evaluation results for local thermal environments as described in Section 6.4. In cases with air supplied from desk-mounted diffusers, the air temperature at floor level increased and more uniform temperature was created around the human body, compared with that in cases with air supplied from floor level only. This is beneficial to alleviate and avoid the "cold-feet" phenomenon, which is especially obvious in Case 2 as shown in Figure 6.3, when all of the air is supplied from desk-mounted diffusers. On the other hand, the stratification temperatures in the upper zone were very similar, which was desirable for energy efficiency.

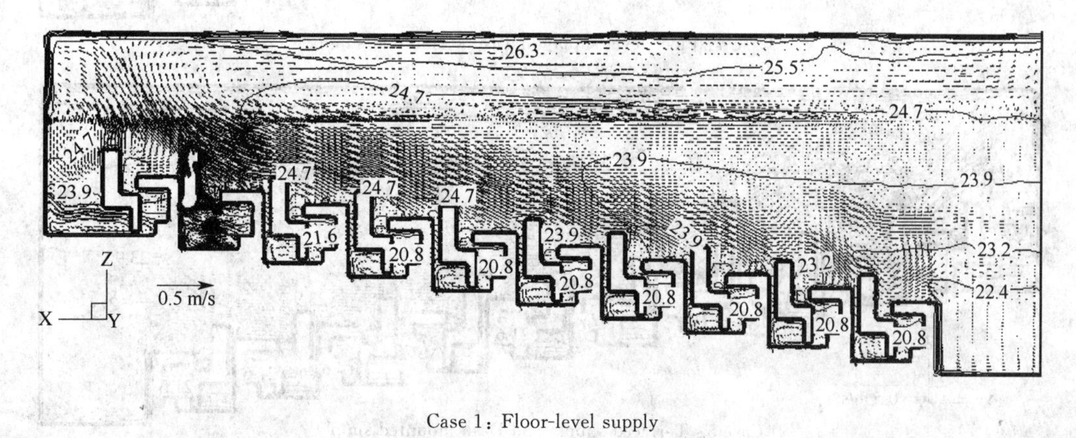

Case 1: Floor-level supply

Figure 6.3 Velocity and temperature distribution with different air supply methods

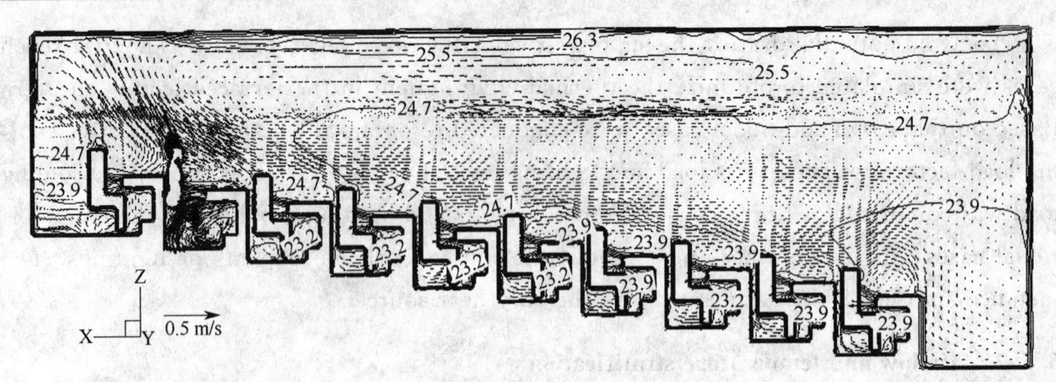

Case 2: Desk-mounted supply

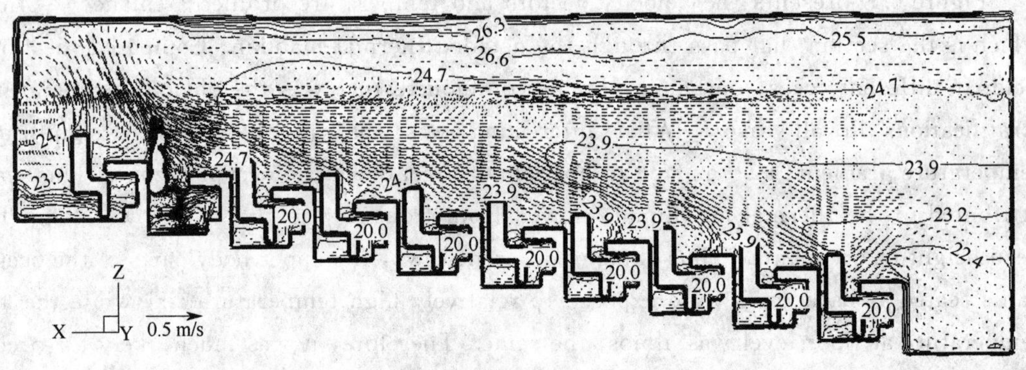

Case 3: Terraced supply

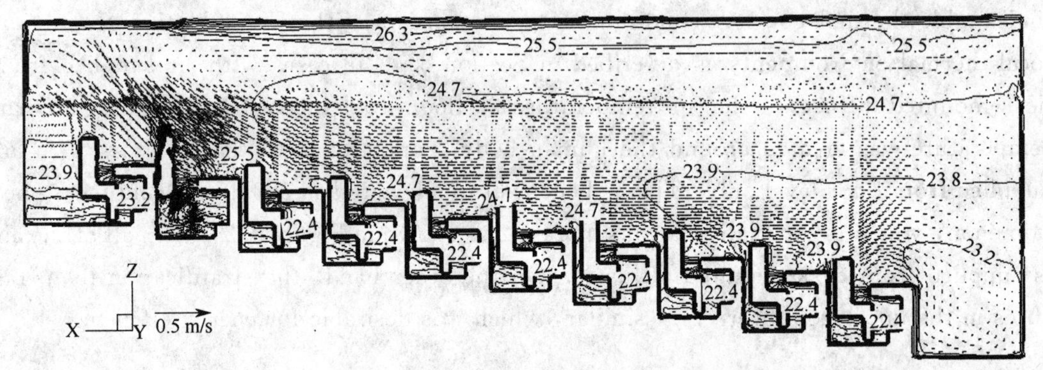

Case 4: Floor-level supply and Desk-mounted supply

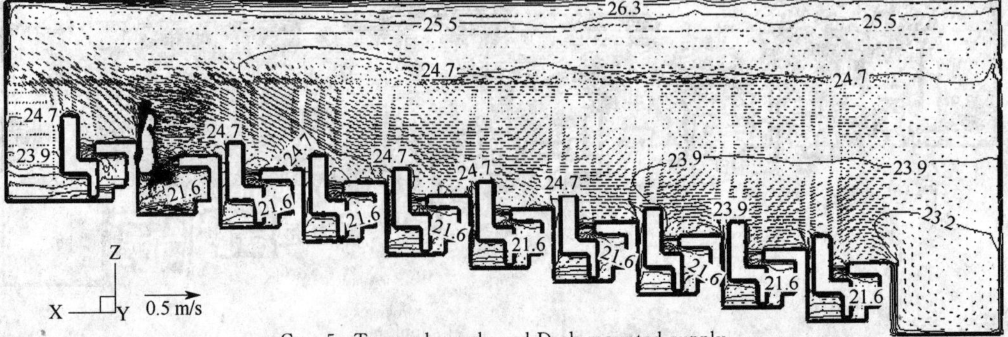

Case 5: Terraced supply and Desk-mounted supply

Figure 6.3 Velocity and temperature distribution with different air supply methods (continue)

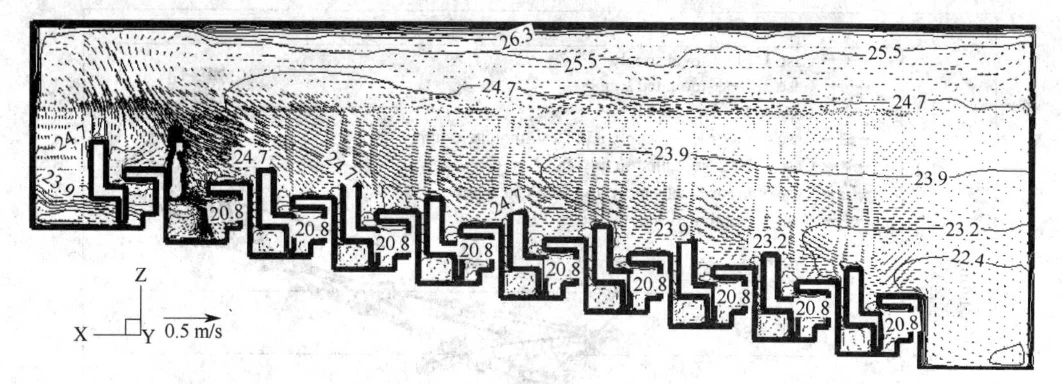

Case 6: Floor-level supply and Terraced supply

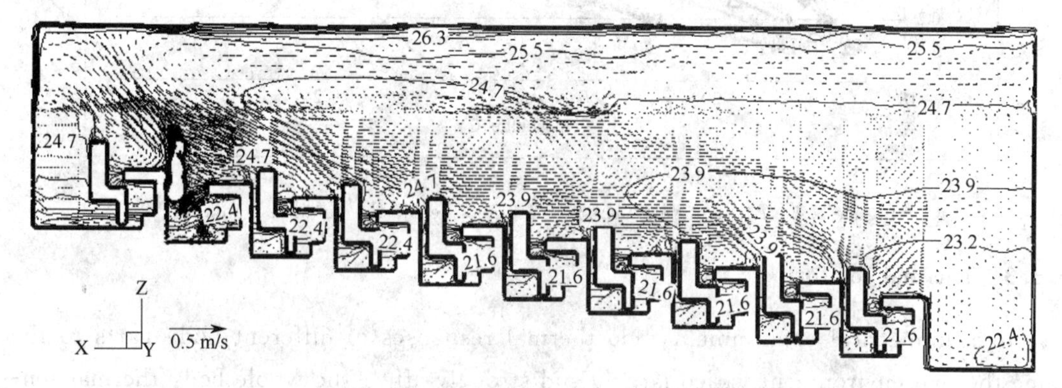

Case 7: Floor-level supply, Terraced supply and Desk-mounted supply

Figure 6.3 Velocity and temperature distribution with different air supply methods (continue)

Figure 6.4 presents the temperature stratifications at the position that 0.15m away from right side of the numerical thermal manikin. It was also clearly shown that when all of the air was supplied from floor level, the feet of occupants were directly exposed to the cold flow draft, which may cause un-comfortable feelings. The thermal comfort for the feet was largely improved when there was air supplied from desk-mounted diffusers. When encountering the chest of occupants, the cold supply air from desk-mounted diffusers separated into two airstreams. Part of the supply air was entrained by the uprising thermal plume of the occupant and the rest air flowed downward due to higher air density than surrounding warm air. Thus, the temperature difference between head and ankle levels was reduced, which was beneficial to improving the thermal environment. This virtue was especially apparent in Case 2, when all of the air was supplied from desk-mounted diffusers. The air temperature at 0.1m in Case 2 was about 4.5℃ higher than that in Case 1 or Case 3. When supplying half of the total air from the desk-mounted grilles as that in Case 4, 5 and 7, the ankle level air temperature also increased. Temperature gradients in the occupied zone in these cases were obvious smaller than those in Case 1, 3 and 6, when all the air was supplied from floor level, and thus better thermal comfort may be achieved.

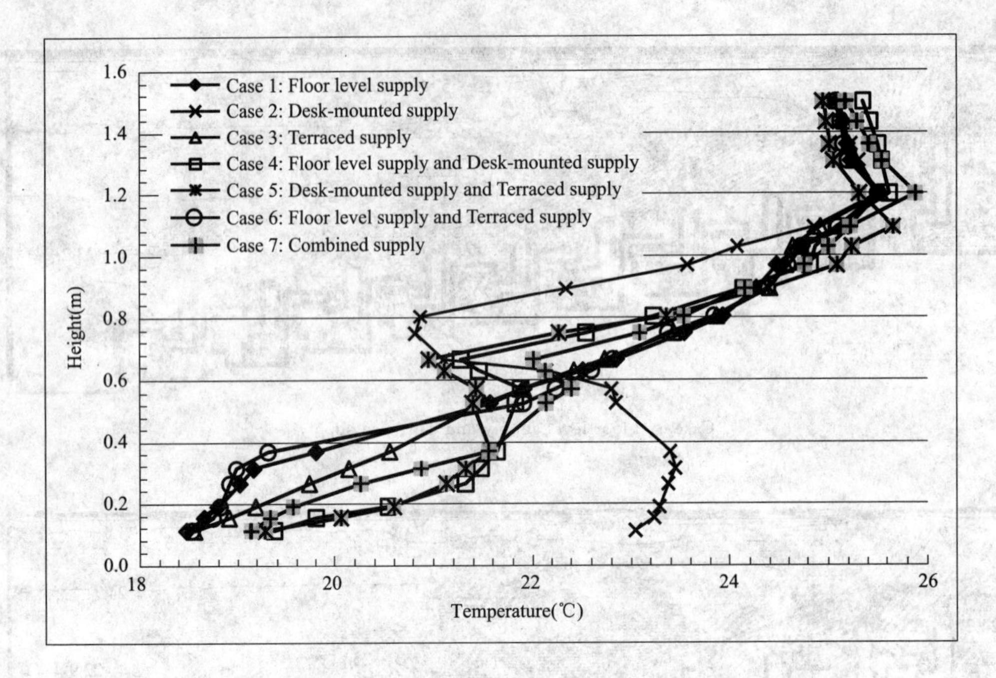

Figure 6. 4 Air Temperature stratifications beside the detailed manikin

6. 5. 3 Thermal comfort

In asymmetrical environment, the thermal responses of different body parts to the same thermal environment varied largely and strongly affect the whole body thermal sensation. The whole body thermal sensation also influences the local body thermal sensations in turn. Hence thermal comfort evaluation in asymmetrical environment is more complicated than that in uniform environment. Figure 6. 5 presents thermal sensation and comfort for local body parts and whole body of the numerical thermal manikin evaluated by using the UCB thermal comfort model. It indicated that the local thermal sensations were most uniform in Case 7, when the air was supplied from three locations simultaneously and relatively uniform thermal environment in occupied zone was formed. In Cases 1, 3 and 6, when the conditioned air was respectively supplied only from the floor-level, from terrace, and from these two locations together, local overcooling occurred for lower body segments.

When part of the air was supplied from desk-mounted diffusers, thermal sensations in those body parts were improved as indicated in Cases 4 and 5. Thermal sensations of the upper body parts were not as sensitive as that for the lower body parts to different air distribution methods. The main reason was that thermal sensations for the upper body were influenced not only by the local air flow but also strongly by the whole space air flow, especially by the thermal air plumes from the front-row occupants. Local and whole body thermal comforts were acceptable in all the seven cases, and the variations in thermal comfort could be attributed to the change of thermal sensations. In Case 2, when the cooling air

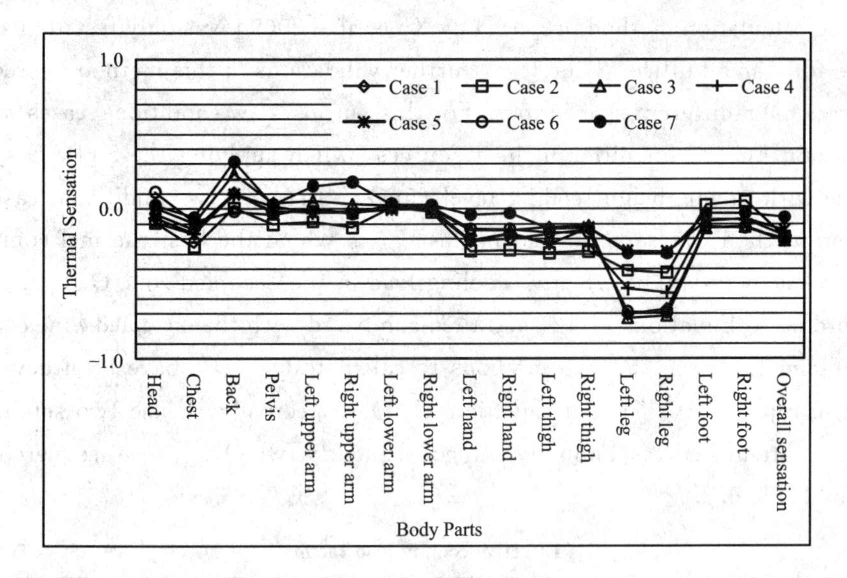

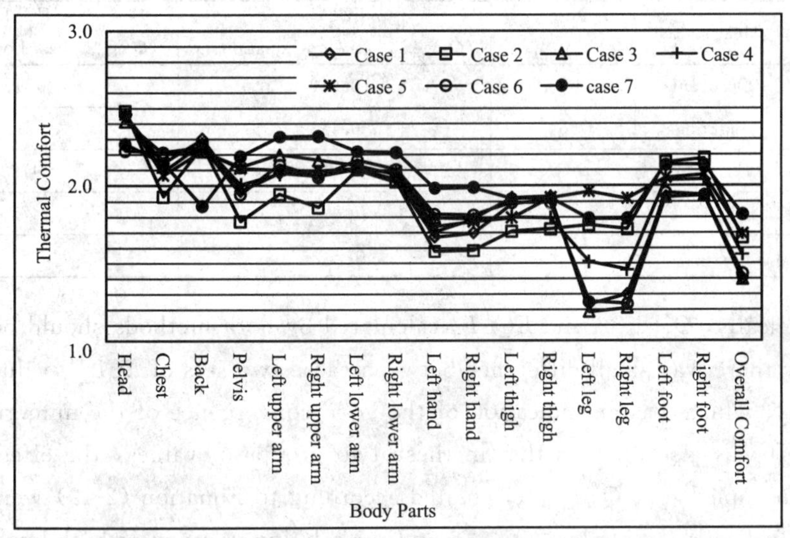

Figure 6.5 Thermal sensation and comfort for local body parts and whole body with different air distribution methods

was supplied directly to chest and pelvis of occupants, no obvious cold draft was caused. That was because the clothing insulation at chest and pelvis makes the human body less sensitive to the cooling air. In Cases 1, 3 and 6, with no air was supplied from desk-mounted grilles, the thermal comfort was worse compared with that in other cases. It appears that better thermal comfort can be achieved by supplying air from chest height level than by supplying air from floor level only.

6.5.4 Occupied zone cooling load calculations

The key point in the designing of a STRAD system is to determine a suitable supply airflow rate according to the occupied zone cooling load. The CFD-based occupied zone cooling load calculation method proposed by Xu et al. [2009] was only tested in simulation conditions for a small office. Therefore, further validations of this method are required for the applications in different space types. For this purpose, two additional cases were simulated to obtain $ECLF_i$ for different heat sources, when applying the STRAD systems in large space with middle-height ceiling level (5m). In these two additional cases the air distribution method was same to that in Case 7, in which the best thermal comfort environment was achieved. The effective cooling load in the occupied zone $Q'_{occupied}$ was calculated according to Equations (2-12) and then compared with the occupied zone cooling load $Q_{occupied}$ obtained in Case 7. Two methods to calculate $Q_{i\text{-}occupied}$ have been developed, as presented in Equation (2-8) and Equation (5-1) respectively, hence two sets of $ECLF_i$ values were obtained correspondingly and consequently two $Q'_{occupied}$ values were calculated as listed in Table 6.2.

Effective cooling load factors **Table 6.2**

Cases	Heat source	$Q_{i\text{-}space}$ (W)	$ECLF_i$ ($Q_{i\text{-}occupied}$ based on Equation 5)	$ECLF_i$ ($Q_{i\text{-}occupied}$ based on Equation 8)
Case 8	Occupants	4554	0.912	0.916
Case 9	Lightings	764	0.713	0.721
$Q'_{occupied}$(W)			4698	4722
$Q_{occupied}$ in Case 7(W)			4742	4750
$\lvert Q'_{occupied}-Q_{occupied}\rvert/Q_{occupied}$			0.93%	0.59%

Theoretically, $Q_{i\text{-}occupied}$ and $ECLF_i$ calculated by both methods should be the same. But actually there was small difference between these two sets of $ECLF_i$ values as shown in Table 6.2, which was an indication of the good convergence of the numerical solution procedure. It was also obvious that in these two groups of values, the effective cooling load in the occupied zone $Q'_{occupied}$ calculated according to Equation (2-12) were very close to the occupied zone cooling load $Q_{occupied}$ and the relative errors were both lower than 1%. This confirmed that the occupied zone cooling load calculation method proposed by Xu et al. [2009] was applicable for the conditions simulated in this chapter, i.e. for large-space lecture theatre conditioned by the STRAD system with separate location of return and exhaust grilles.

6.5.5 Energy saving

As shown in Equation (5-1), the divisions of each heat sources in convection and radiation portions are directly influence the occupied zone cooling load. Table 6.3 presents the radiation heat distributions of each heat sources in different conditions. The occupants and lightings were the only two heat sources in the space. The former is dominant and takes about 85% of the total heat. It could be found that the radiative fraction of the heat gain from occupants was varied from 45.9% to 51.6% among different cases, while from lightings this value was about 85% and relatively stable. Most of the radiation heat was distributed into the occupied zone and absorbed by the floor and desk surfaces as shown in the table, which should be considered into the occupied zone cooling load. Whilst the rest of radiation heat gain released to the un-occupied zone and the convection heat gain from lighting should be excluded in determining the required supply air flow rate.

The radiation heat distribution **Table 6.3**

	Radiative Fraction		Radiation heat gain				
	Occupants	Lighting	Floor	Desk	Side wall (Occupied zone)	Side wall (Un-occupied zone)	Ceiling
Case 1	51.5%	84.7%	33.2%	41.4%	11.0%	9.8%	4.6%
Case 2	45.9%	86.6%	29.4%	43.1%	11.5%	11.9%	4.1%
Case 3	51.6%	85.5%	35.4%	39.8%	11.0%	9.4%	4.4%
Case 4	49.6%	86.3%	31.5%	40.8%	11.8%	11.2%	4.7%
Case 5	49.9%	86.9%	32.1%	40.5%	11.6%	11.2%	4.6%
Case 6	50.8%	85.1%	33.0%	41.4%	10.9%	10.1%	4.6%
Case 7	50.1%	86.0%	32.4%	40.9%	11.4%	10.6%	4.7%

As illustrated in Section 4.1, there are two main energy-saving potentials that can be available for a STRAD system. First, in a STRAD system, only the heat gains contributed to the occupied zone $Q_{occupied}$ (W) needs to be considered when determining the supply air flow rate. Second, by splitting location of return and exhaust grilles in STRAD system, the cooling coil load is able to be further reduced. The cooling coil load reductions ΔQ_{coil} in different cases were calculated according to Equation (4-9) and presented in Table 6.4. $t_{occupied}$ (℃) is the mass-weight average air temperature at head level in the occupied zone, which was obtained from the post-process of CFD simulation results.

Energy saving potentials with different air distribution methods **Table 6.4**

	t_r	t_e	$t_{occupied}$	ΔQ_{coil}	$\Delta Q_{coil}/Q_{Space}$	$(Q_{Space}-Q_{occupied})/Q_{Space}$
Case 1	25.4	25.9	24.0	830	15.6%	10.5%
Case 2	25.4	25.9	24.4	655	12.3%	10.3%
Case 3	25.3	25.9	24.0	830	15.6%	10.3%
Case 4	25.2	25.8	24.2	699	13.2%	11.0%
Case 5	25.3	25.9	24.2	742	14.0%	10.9%
Case 6	25.3	26.0	24	874	16.5%	10.6%
Case 7	25.3	25.9	24.1	786	14.8%	10.6%

In all the cases, the exhaust air temperature was higher than the return air temperature by about 0.5℃ to 0.7℃, which resulted in the extra energy savings. When calculating ΔQ_{coil} according to Equation (4-9), the t_{set} was substituted by the Mass-weighted average air temperature at head level in the occupied zone $t_{occupied}$. The cooling coil load reduction varied from 12.3% to 16.5% of the space cooling load, depending on the air supply methods. Table 6.4 also reveals that the space cooling reduction, which defined as $Q_{Space} - Q_{occupied}$, is basically stable and about 11% of the space cooling load. This counters to the expectation that the energy saving of the cooling coil should be less than the space cooling load reduction. That mainly caused by the buoyancy-induced flows in the upper slant plane boundary of the occupied zone. Close examination of velocities vectors on the occupied zone boundary plane as illustrated in Figure 6.6 revealed that downward recirculation occurs and that the temperature at the back rows is about 1.4℃ higher than that at the front rows. The air at the back rows with higher temperature tended to flow to the ceiling level exhaust grilles directly, therefore resulted in direct exhausting of thermal load originating from the occupied zone and reducing of the cooling coil load.

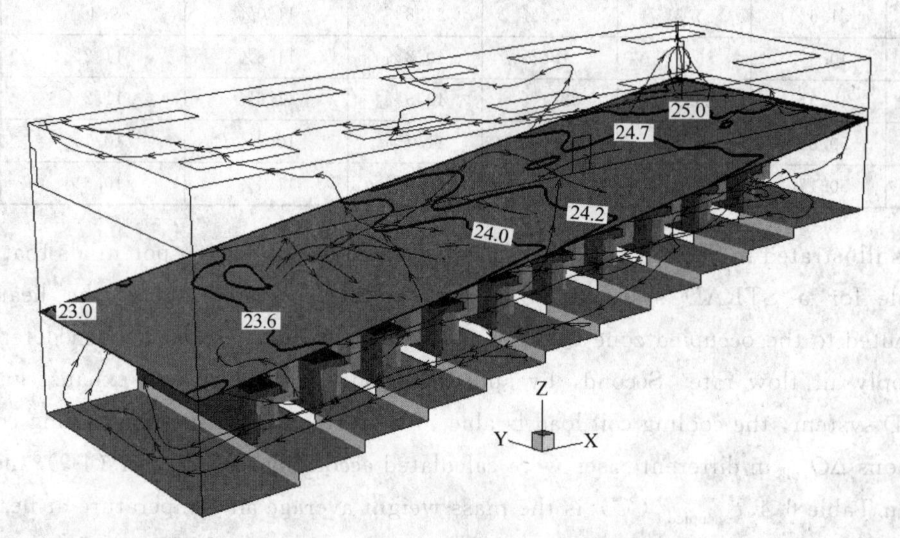

Figure 6.6 Streamlines and temperature distribution in the upper boundary of the occupied zone

6.6 Different heights of return grilles

In Section 6.5 different air supply methods to realize stratified air distribution in a lecture theatre were studied for the purpose of optimizing the thermal environment and maximizing the energy saving potentials. In these simulation cases, the return grilles were located at the middle level of surrounding walls with a fixed height. However, the height of return grilles has been certified to strongly affect the performance of STRAD systems, when applied in the small office in Chapter 5. Therefore, the impact of the return grilles height on the performance of STRAD systems, when applied in the large lecture theatre, is worth to be studied.

6.6.1 Simulation cases description

It has been demonstrated in Section 6.5 that best thermal comfort was achieved in Cases 7, when the air supplied from three locations simultaneously. Therefore, the air distribution method in Case 7 was adopted in this section. All the parameters were set same to that in Case 7 except for the height of return grilles. Four return grilles are still distributed at surrounding walls, but the heights are varied in different cases as indicated in Figure 6.7. Seven cases, as listed in Table 6.5, are simulated to investigate the influence of return grilles height on the performance of the STRAD system.

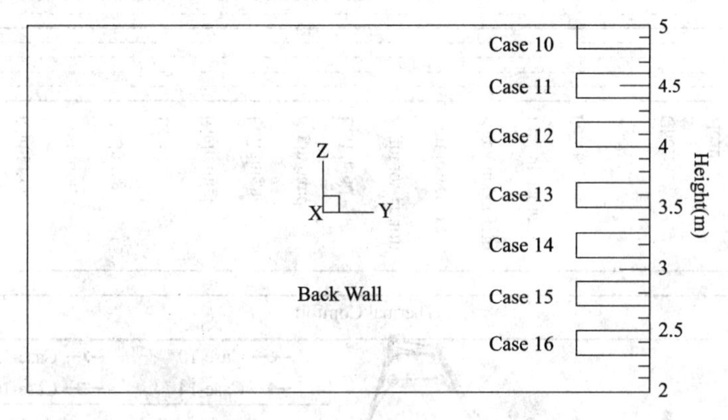

Figure 6.7 Simulation cases with different heights of return grilles

Simulation cases with different heights of return grilles **Table 6.5**

Simulation Cases	Case 10	Case 11	Case 12	Case 13	Case 14	Case 15	Case 16
Height of return inlet* (m)	4.8	4.4	4	3.5	3.1	2.7	2.3

Height of return inlet* : The height is from floor level to the bottom edge of the return grille.

6.6.2 Thermal comfort

Figure 6.8 presents the thermal sensation and thermal comfort for the numerical thermal manikin located at ninth row in different cases, evaluated by using the UCB thermal comfort model. It indicated that as the height of return grilles decreased from ceiling level (Case 10) to the lower occupied zone (Case 16), the thermal comfort for the thermal manikin was affected, which was especially obvious at head level. The main reason was that the numerical thermal manikin was located close to the return grille at back wall. When the height of return grille located at back wall decreased from ceiling level to floor level, the induced function of the return grille led portion of the cold supply air return to the AHU directly, and thus the head experienced less cold airstreams. Whilst the head level was always exposed to the thermal plumes flowed from front rows of occupants and moved to the warmer side as the height of return grille decreased. Therefore, the local thermal environment for the head level was deteriorative when the height of return grilles decreased from ceiling level to floor level. This is identical with the findings in Chapter 5 for a small office that the enhanced influence of induced function from return grille leads to large temperature increasing at head level. However, the space of the lecture theatre was much larger than the small office studied in Chapter 5 and most of the return grilles were

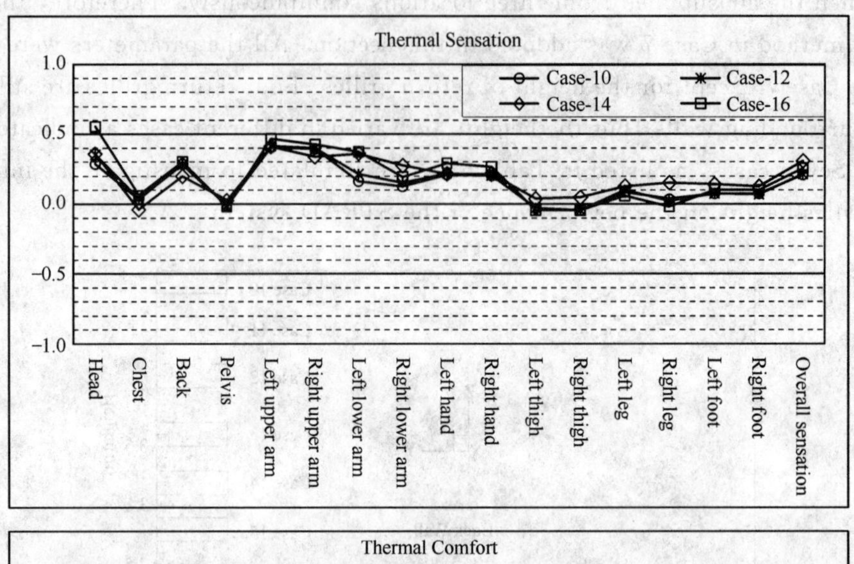

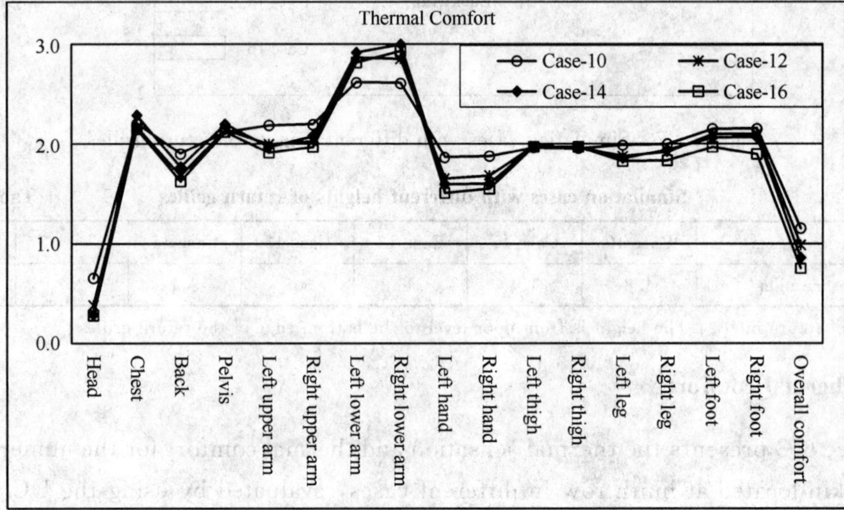

Figure 6. 8 Thermal sensation and comfort for local body parts and whole body with different heights of return grilles

located far away from the occupants. Thus, the thermal environment was deteriorative slightly and not as intense as that in Chapter 5 for the small office. Figure 6. 9 presents the predicted percent dissatisfied due to draft (PD) at Plane Z=2m in Case 15, which is the bottom height of return grilles. It was obvious that the local draft risk caused by the momentum flux of return grilles could be ignored excepted for back rows, where the return grille was located close to the occupants. The overall thermal comforts in different cases were all in accepted range. Thus, it was supposed that when applying a STRAD system in a large space, satisfied thermal environment could be maintained as locating the return grilles at lower occupied zone. To weaken the impact of induced function from return grilles to the thermal comfort of occupants, it is recommended to locate the return grilles far away from occupants.

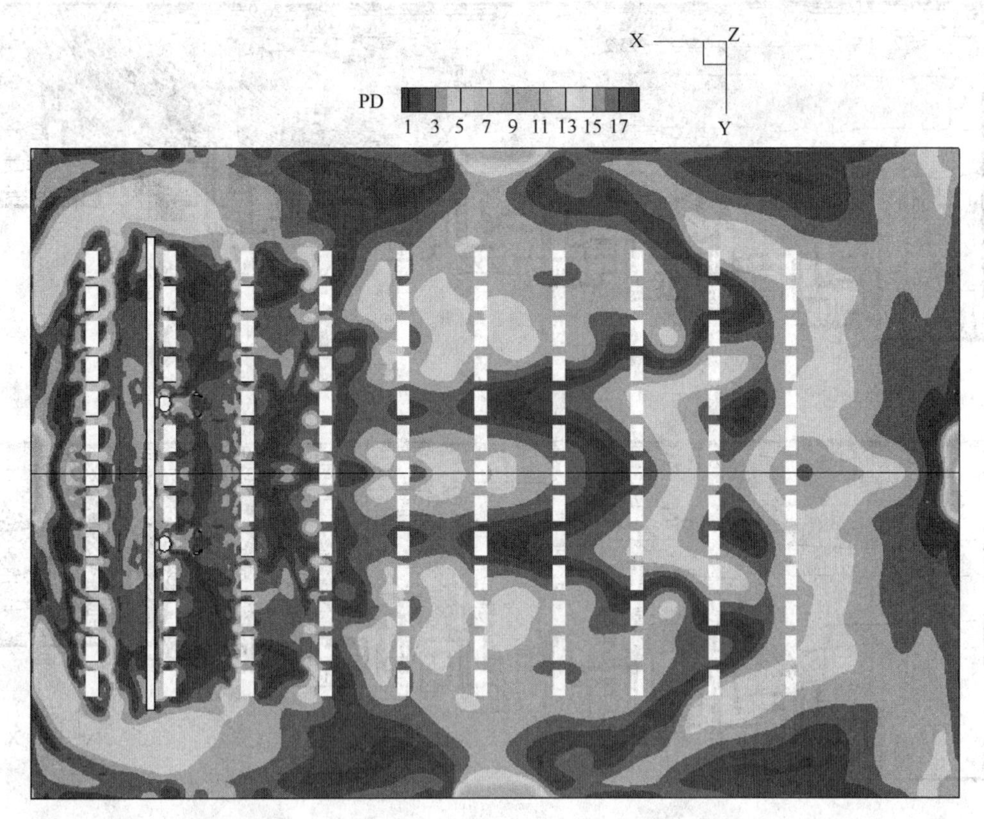

Figure 6.9 Values of predicted percent dissatisfied due to draft at Plane Z=2 m in Case 15

6.6.3 Energy saving

Figure 6.10 presents the velocity and temperature distributions at Plane Y=4m for three cases with different heights of return grilles. It indicated that decreasing the height of return grilles has slightly impact on the overall air distribution in the occupied region. However, as the height of return grilles declined from ceiling level to floor level, the air temperature at upper zone increased rapidly, which was especially obvious for the front part. That meant much more convective heat was transferred to the upper zone and possible to be exhausted from ceiling diffusers directly.

The energy savings for the cooling coil in different cases were calculated according to Equation (4-9) and listed in Table 6.6. As the return grilles height decreased from upper zone to the occupide zone, the exhaust air temperature increased, which indicated much more heat was dischaged from the exhaust inlets and exclued from the cooling coil load. Corresponding to that, the energy saving of the cooling coil increased from 14.8% to 21.4% of the space cooling load. The energy savings for the STRAD system, when applied in a large lecture theatre with higher ceiling, was more prominent than that applied in a small office. This is consists with the conclusion that room with higher ceiling is beneficial for improving the energy efficiency of displacement ventilation systems [Hashimoto and Yoneda 2009].

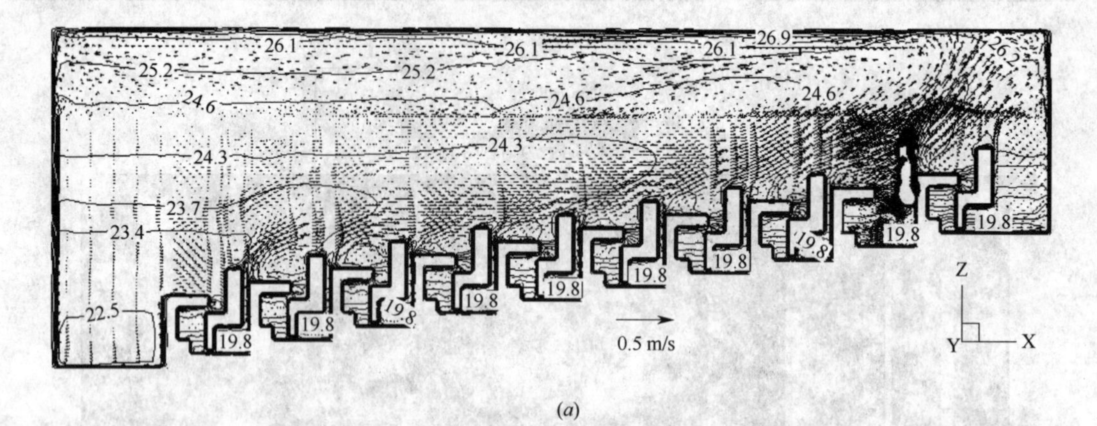

(*a*)

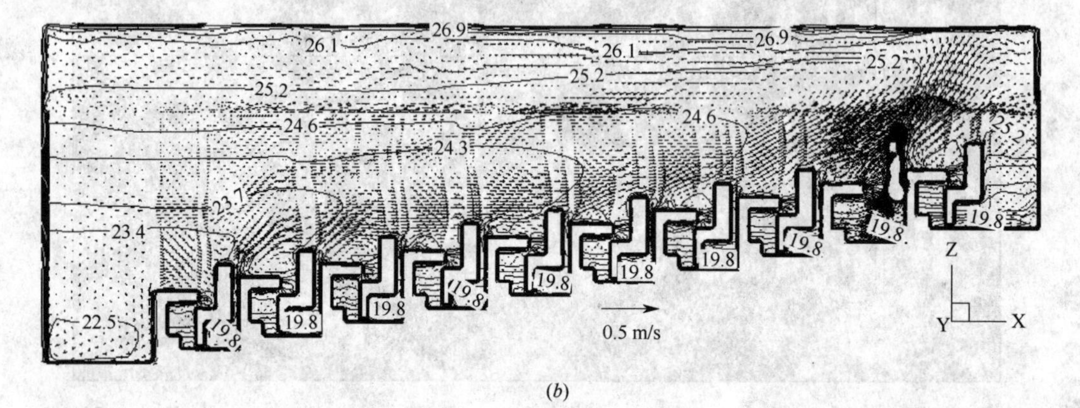

(*b*)

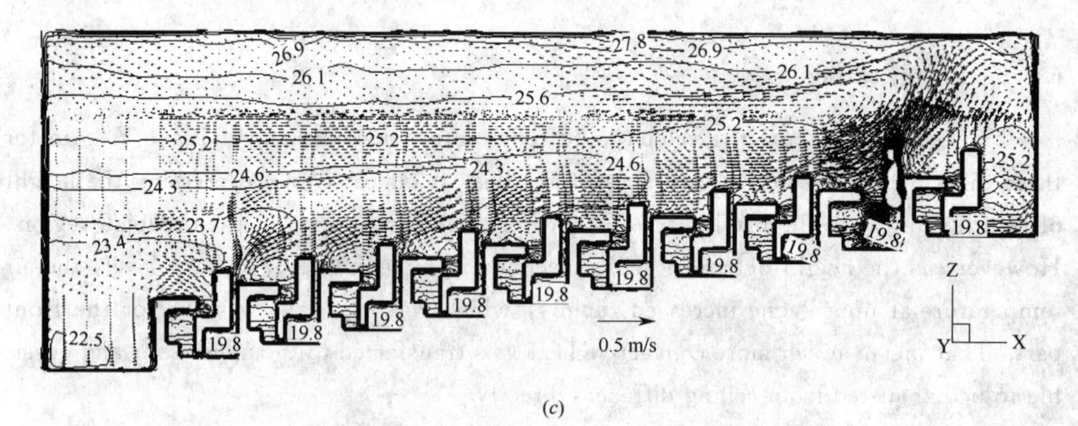

(*c*)

Figure 6.10 Velocity and temperature distribution at Plane Y=4m in different cases

(*a*) Case 10; (*b*) Case 13; (*c*) Case 16

Energy-savings for cooling coil **Table 6.6**

Cases	Height of return grilles (m)	t_r(℃)	t_e(℃)	ΔQ_{coil}(W)	$\Delta Q_{coil}/Q_{Space}$
Case 10	4.8	25.9	25.8	786	14.8%
Case 11	4.4	25.2	25.9	830	15.6%
Case 12	4	25	25.9	830	15.6%

contiune table

Cases	Height of return grilles (m)	t_r(℃)	t_e(℃)	ΔQ_{coil}(W)	$\Delta Q_{coil}/Q_{Space}$
Case 13	3.5	24.8	26.1	917	17.3%
Case 14	3.1	24.6	26.2	961	18.1%
Case 15	2.7	24.3	26.4	1048	19.7%
Case 16	2.3	23.9	26.6	1136	21.4%

6.6.4 Effective cooling load factors

The effective cooling load factors of different heat sources in cases with different heights of return grilles were calculated. Since occupants and lightings were the only two heat sources in the space, there totally 14 additional cases were simulated, corresponding to the seven return grilles heights as listed in Table 6.6. The CFD settings for these 14 cases were same to the corresponding cases with the same return grilles height as listed in Table 6.6, except the boundary conditions for the heat sources. In each of these 14 cases, one of the two heat sources was maintained, while the other one was turned off. For instance, when calculating the effective cooling load factor of occupants with a return grilles height of 4.8m, the CFD settings in the simulation case was same to that in Case 10 but with turned off lightings.

Table 6.7 presents the effective cooling load factors for each heat source with different height of return grilles. The effective cooling load factor for the occupants increased slightly from 0.83 to 0.90 as the height of return grilles was decreased from 4.8m to 2.3m, while the value for the lightings increased little more apparent, from 0.70 to 0.82. Corresponding to that, the calculated effective occupied zone cooling load $Q'_{occupied}$ increased as well, which was in line with the variation of the occupied zone cooling load $Q_{occupied}$. The divergences between $Q'_{occupied}$ and $Q_{occupied}$ in all conditions were less than 10.0%, which manifested the occupied zone cooling load calculation method proposed by Xu et al. [2009] was effective for STRAD systems applied in a large lecture theatre with similar configurations.

Effective cooling load factors for each heat sources **Table 6.7**

$ECLF_i$	Q_i(W)	Return grilles height						
		4.8m	4.4m	4m	3.5m	3.1m	2.7m	2.3m
Occupants	4546	0.83	0.84	0.84	0.85	0.86	0.88	0.90
Lightings	764	0.70	0.72	0.76	0.78	0.79	0.81	0.82
$Q'_{occupied}$(W)		4308	4369	4399	4460	4513	4619	4718
$Q_{occupied}$(W)		4695	4712	4747	4766	4781	4795	4814
$(Q'_{occupied}-Q_{occupied})/Q_{occupied}$		−7.3%	−6.5%	−6.5%	−5.8%	−5.0%	−3.3%	−1.8%

However, it seems contradictory that both the occupied zone cooling load and the cooling coil load reduction increased as the height of return grilles decreased. One main reason is that the temperature of reversed air, from the upper zone to the occupied zone, increased largely when the height of return grilles decreased from ceiling level to floor level. Figure 6.11 compares the temperature and local air velocity distributions at Plane Y=4m

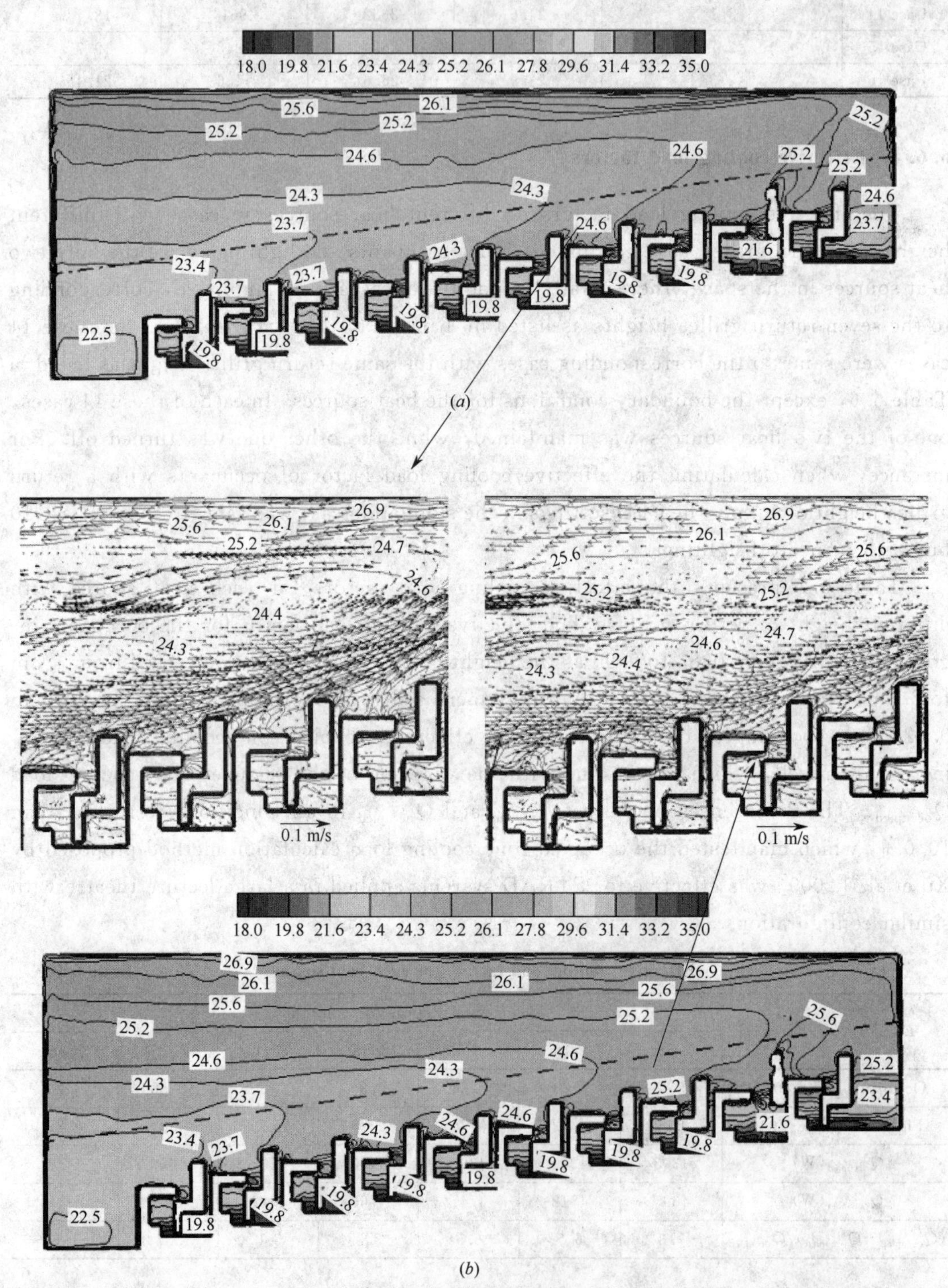

Figure 6.11 Temperature and velocity distribution at Plane Y=4m in Case 11 and Case 15

(*a*) Case 11; (*b*) Case 15

in Case 11 and Case 15. The red dashed line is the upper boundary of the occupied zone. It clearly shown that the reversed air temperature in Case 15 is much higher than that in Case 11, which means more heat is entrained from the upper zone to the occupied zone. Thus, the occupied zone cooling load in Case 15 is larger than that in Case 11.

On the other hand, as clarified in Section 3.3, the cooling coil load reduction was associated with but differential from the occupied zone cooling load. Not all of the heat contained in the reversed flow is contributed to the cooling coil load. The warm air from occupants is possible to be exhaust directly and excluded from the cooling coil load, which is similar to the findings as shown in Figure 6.6. Therefore, as the height of return grille decreased, the occupied zone cooling load increased, and the cooling coil load reduction increased simultaneously, which indicates that portion of the heat from occupied zone is discharged directly from exhaust grilles and excluded from the cooling coil load.

6.7 Discussion and Summary

The heat gain from lightings can be divided into short-wave radiative fraction, long-wave radiative fraction and convective fraction [ASHRAE 2009]. The values of radiative fraction presented in ASHRAER Standard 2009 were obtained in the situations that conditioned air was returned through the ceiling grilles or luminaire side slots. In present simulations, we assumed that the enclosure or furniture surface has the same absorption coefficients for radiation of different wavebands. Also, the luminaire side slots were not considered, which weaken the convection heat exchange between lightings and exhaust air. Hence, the simulation results revealed the radiative fractions of lightings in Section 6.5 were about 85%, which might be higher-biased.

In a large lecture theatre with terraced floor, different designs to realize thermal stratification were studied in Section 6.5. The simulation results showed that the air temperature around the upper body parts at back rows was somewhat higher than that at front rows because of the upward thermal air plumes flowing along the terraced seats. The results also revealed that the distributed air supply nozzles were able to realize temperature stratification and thermal comfort at the same time. Supplying air from desk-mounted grilles achieved better thermal sensations of lower body parts in comparison with supplying air from floor level only, which is beneficial for the overall thermal comfort of occupants. Thermal environment was optimized by supplying air from the most distributed supply locations, namely from the three locations of supply diffusers simultaneously. Further energy saving was obtained by splitting the locations of return and exhaust grilles. The energy saving was maximized and up to the value of 16.5% of the space cooling load was saved for the cooling coil when supplying air from floor-level and terrace diffusers simultaneously .

The influences of return grilles height on the ventilation performance of the stratified air distribution system in the large lecture theatre were also investigated. The results indicated that decreasing the height of return grille locations from ceiling level (4. 8m) to floor level (2. 3m), the thermal environment deteriorative slightly. Satisfied thermal environment could be realized in a large space when locating the return grilles at lower occupied zone in a STRAD system. However, to weaken the negative impact of induced function of return grille to the thermal comfort of occupants, it is recommended that the return grilles are not located too close to the occupants. Decreasing the height of return grille locations increased the exhaust air temperature, and much more heat was dischaged from the exhaust grilles directly. Corresponding to that, the energy saving of the cooling coil increased largely. Meanwhile, there was also a the temperature increase of reversed air flowing from upper zone to the occupied zone. Thus, the occupied zone cooling load $Q_{occupied}$ increased. However, the reversed flow is varied case by case and difficult to be predicted. The occupied zone cooling load calculation method proposed by Xu et al. [2009] was validated for the application of STRAD systems in a lecture theatre with large space and high ceiling (5m) . The divergence between $Q'_{occupied}$ and $Q_{occupied}$ was small enough to verify the effectiveness of this method. The effective cooling load factors for each heat sourece in different conditiones were calculated based on 14 additional simulation cases.

Chapter 7 Alternative stratified air distribution designs in a terminal building

7.1 Introduction

Modern buildings with high ceilings and large floor areas, such as airport terminal buildings and international conference halls, usually have large glazed facades because of facades' architectural aesthetics and advantage of introducing daylight into indoor spaces [Gordon 2008]. However, large areas of fenestration admit large solar heat gains which contribute to the building's thermal load, and occupants may also suffer from glare due to direct solar radiation and feel overheated. In addition, the internal heat gains in these buildings are always high due to high occupant densities and illumination intensities, making the energy consumption of such buildings considerable [Parker et al. 2011]. A survey of 29 Greek airports revealed that the average annual total energy consumption at the airport terminals was $234kWh/m^2$, approximately 50% of which was used for HVAC systems [Balaras et al. 2003]. The energy saving potential of the HVAC systems in all 29 airports was evaluated, and the results indicated that considerable energy could be saved through a number of different energy conservation measures. For example, a 16%~32% reduction in cooling energy consumption could be achieved by using hybrid cooling systems, such as ceiling fans, which would maintain and improve indoor environmental quality at the same time [Balaras et al. 2003]. On the other hand, the occupied zone accounted for only a very small portion of the total available space in buildings with high ceilings. Thus, an energy-conserving design concept that conditions only the first 2m above the occupied zone by utilizing a displacement ventilation system was proposed by Simmonds [1996] for a terminal building. The application of a hybrid conditioning system with a variable-volume displacement conditioning system and a radiant cooled floor in terminal buildings also demonstrated the possibility of considerably reducing energy consumption [Simmonds et al. 1999].

The thermal buoyancy arising from vertical temperature differences plays an important role in determining the indoor air flow of terminal buildings with high ceilings. The indoor thermal environments in these buildings are subject to rapid deterioration by the radiant heat or outer thermal conditions [Kim et al. 2001]. Different air distribution designs were investigated to optimize the thermal environment in a train station building with a high ceiling [Li et al. 2009]. The numerical results indicated that satisfactory thermal comfort in the occupied region was realized by using stratified air distribution design and supplying air at mid-height horizontally. Han and Gu [2008] numerically studied the thermal environment in the 3rd terminal building of the Beijing Capital International Airport, in which the air was horizontally supplied from integrated air conditioning units named binnacles with a height approximately 3m above floor level. The results revealed that a satisfactory thermal environment was achieved in the occupied zone.

Such limited research on the application of STRAD systems in large spaces offers little insight into their energy-saving aspects. Thus, this chapter numerically studies two typical and practical air distribution designs that can be used to realize thermal stratification in a hypothetical terminal building, with our newly defined energy reduction index applied. The air is supplied horizontally at floor level or mid-height (approximately 2m above floor level) . The effect of diffuser locations on the thermal environment and energy consumption are investigated, and the impact of solar radiation on the performance of the STRAD system is also illustrated. The energy savings of the cooling coil under different conditions are evaluated and compared according to the method developed in Chapter 4, and the effective cooling load factors for each heat source are calculated. The results obtained in this chapter can help engineers better understand the air distribution design methods of a STRAD system for large spaces.

7.2 Simulation case descriptions

7.2.1 Hall geometry and air distribution designs

The terminal building is assumed to be rectangular in shape and the facades are constructed with single-layer clear glass. Considering the computational resource limitations, only a middle section of an elongated waiting hall in a terminal building, as shown in Figure 7.1 and with a dimension of 24.9m(W) × 11.2m(L) × 8m(H), is investigated. There are 128 seats provided for occupants waiting for departure. 24lamps are simplified as panels and located at ceiling level, and four skylights are installed on the roof 1m above the ceiling level to take advantage of daylight. Two air conditioning units are located at the center of the space.

This chapter focuses on the study of two typical and practical air distribution designs for the ventilation system of an occupied zone. In the first design, the air is horizontally supplied at floor level at a low velocity and returned at mid-height, a setup that comprises a stratified air distribution. In the second design, the air is horizontally supplied at mid-height at a high velocity and returned at floor level, which is more frequently adopted in practice for large spaces. Diffuser 1 and Diffuser 2, located at the air conditioning units, are the main air diffusers used as either supply or return grilles in the different cases. Diffuser 3 and Diffuser 4 are located on side walls, which is rare in practical applications.

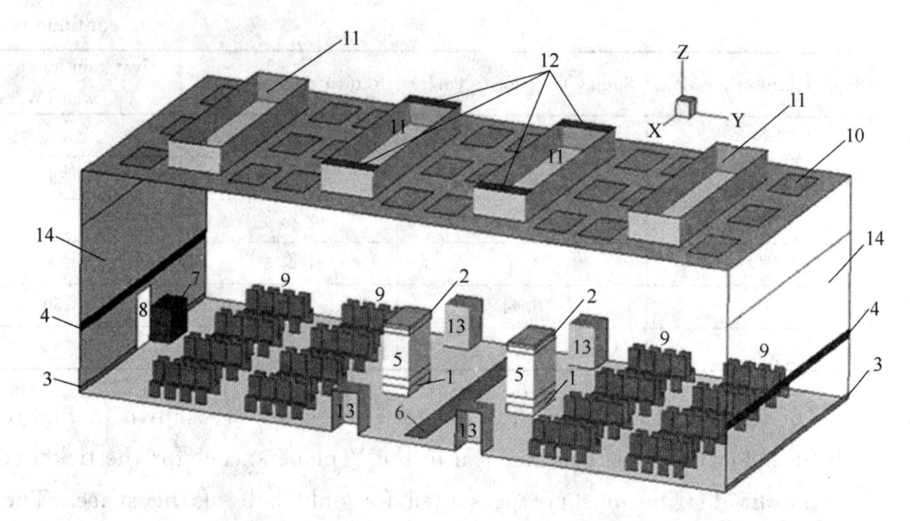

Figure 7.1 Geometry model of the terminal building

1—Diffuser grilles; 2—Diffuser grilles; 3—Diffuser grilles; 4—Diffuser grilles; 5—Air conditioning units; 6—Moving sidewalk; 7—Information desk; 8—Entrance; 9—Occupants; 10—Lighting panels; 11—Skylights; 12—Exhaust grilles; 13—Facilities; 14—Facades

However, to investigate the impact of diffuser locations on ventilation performance, these two diffusers are optionally used as additional return inlets or supply outlets. Diffuser 1 and Diffuser 3 are located at floor level, while Diffuser 2 and Diffuser 4 are located at a height approximately 2m above floor level. It should be emphasized here that the exhaust air grilles are installed near the skylights and completely separated from the return grilles.

7.2.2 Simulation cases and heat gains

There are 12 simulation cases in total, which are listed in Table 7.1. In the first ten cases, the two typical practical air distribution designs are studied and compared, and the effect of the diffuser locations is investigated for each air distribution design. The air distributions in Cases 11 and 12 are the same as those in Cases 9 and 10, respectively. However, the solar radiation of the right-hand glass wall is enhanced in these two cases, and is described in detail in Section 7.2.3.

Different simulation cases **Table 7.1**

Cases	Supply diffuser locations	Supply air velocity (m/s)	Return grille locations	Heat gain for the right side wall (W/m^2)
Case 1	1	0.32	2	29
Case 2	1	0.32	4	29
Case 3	1 & 3	0.31	2	29
Case 4	1 & 3	0.31	4	29
Case 5	2	2.88	1	29
Case 6	2	2.88	3	29

continue table

Cases	Supply diffuser locations	Supply air velocity (m/s)	Return grille locations	Heat gain for the right side wall (W/m^2)
Case 7	2 & 4	2.1	1	29
Case 8	2 & 4	2.1	3	29
Case 9	1 & 3	0.31	2 & 4	29
Case 10	2 & 4	2.1	1 & 3	29
Case 11	1 & 3	0.44	2 & 4	150
Case 12	2 & 4	2.94	1 & 3	150

Schematics of the air distributions for the first ten cases are shown in Figure 7.2. Because the diffuser locations are symmetrical in the Y-plane except for the ticket counter and entrance, the air distributions are presented for only half of the space. The solid arrows indicate the supply air, while the solid arrows with dashed lines indicate the return air. The hollow arrows with dashed lines indicate the exhaust air, which is always discharged at the ceiling level.

The heat sources in the terminal building are summarized in Table 7.2. All of the occupants are assumed to be in a sedentary condition with a sensible heat loss of 70 W/person. The thermal loads of the lamps are set to $16W/m^2$ according to the ASHRAE Standard for a terminal building [2009]. Each facility, such as the electronic screen provided for flight information, is set to 200W, and the power of the moving walkway is

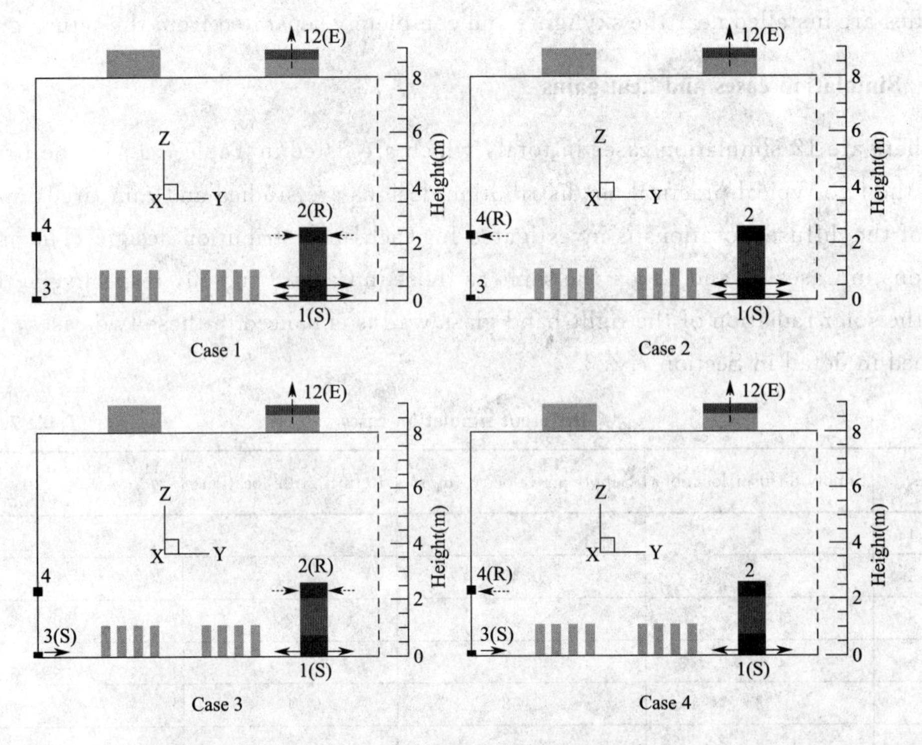

Figure 7.2 Air distributions in different simulation cases

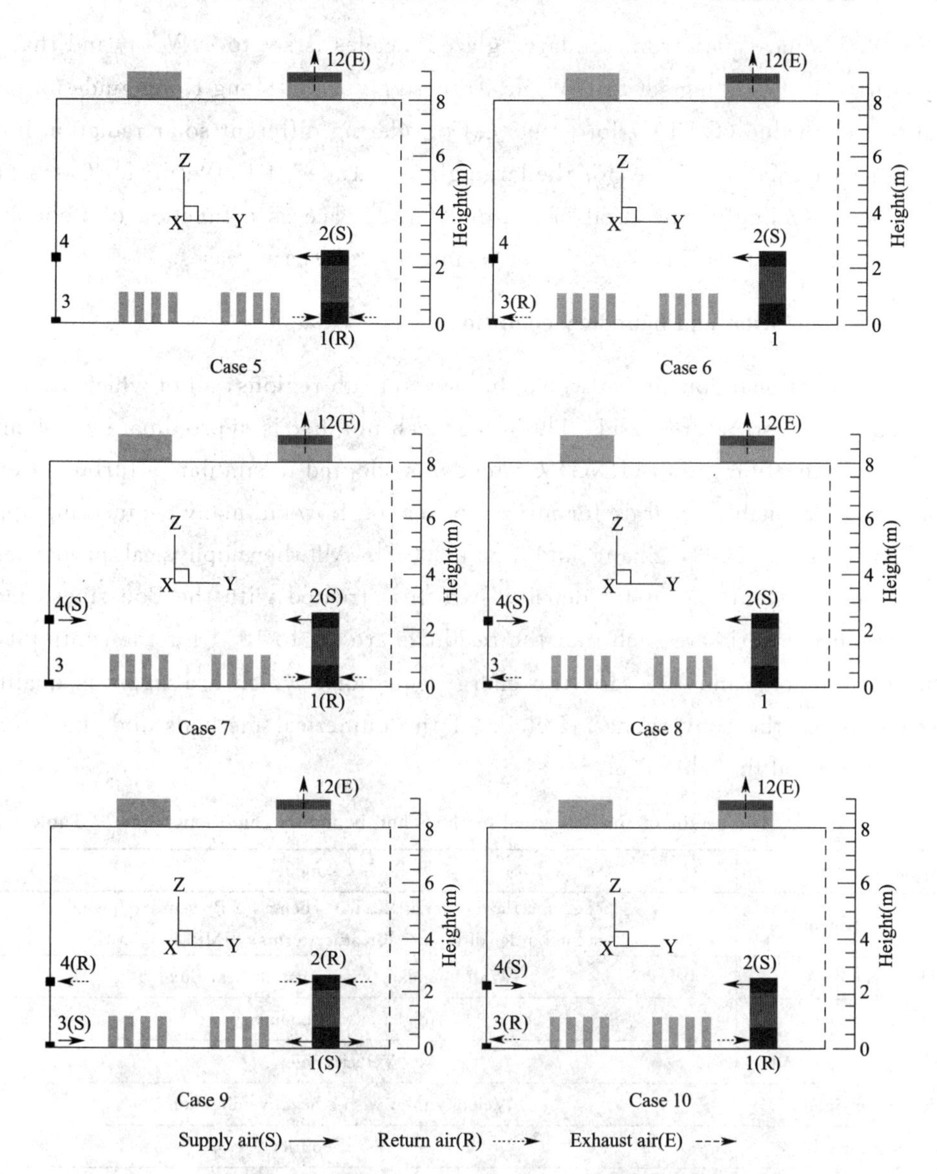

Figure 7.2 Air distributions in different simulation cases (continue)

Heat loads in the terminal building **Table 7.2**

Heat Sources	Heat load
Occupants	70W/person×128
Lightings	16W/m^2×274.4m^2
Facilities	33.9W/m^2×23.6m^2
Left Side Wall	29W/m^2×80m^2
Skylights	19W/m^2×59.2m^2
Ceiling	19W/m^2×157.3m^2
Moving Sidewalk	103W/m^2×7.76m^2
Right Side Wall	29W/m^2(For the first 10 cases) and 150W/m^2(For cases 11 & 12)×84m^2

set to 800W. The heat gain from the large glazed facades is set to $29W/m^2$ and that from the skylights and the ceiling set to $19W/m^2$, according to the Hong Kong guide for overall thermal transfer value (OTTV) for [1995]. Considering different solar radiation intensities, a higher thermal load is set for the large glazed facades at $150W/m^2$ for Cases 11 and Case 12. Correspondingly, the total heat gain of the space as a function of floor area is $85.4W/m^2$ in the first ten cases and $137.5W/m^2$ in the last two cases.

7.2.3 Mesh generation and boundary conditions

The computational domain is divided into several sub-regions, all of which are meshed with a high-quality structured grid. The total mesh number is approximately 1.4 million and the renormalization group (RNG) k-ε model is adopted to simulate a turbulent effect. This model is reasonably accurate for mixed convection flows in many engineering applications [Posner et al. 2003, Zhang and Chen 2007]. All thermophysical properties are assumed to be constant except for density, which is treated with the Boussinesq model. The convergence criteria are such that the residuals are set to 10^{-4} for the continuity and momentum equations and 10^{-7} for the energy equation, which are rigorous qualitative measurements for the convergence. Details of the numerical methods and the boundary conditions are listed in Table 7.3.

The details of the numerical methods and boundary conditions **Table 7.3**

Turbulence Model	RNG k-ε model
Numerical Schemes	Staggered third order PRESTO scheme for Pressure; Upwind second order difference for other terms; SIMPLE algorithm
Heat Sources	Wall boundary with constant heat flux
Floor Surface	Adiabatic wall boundary
Supply air outlet	Velocity inlet
Return air inlet	Velocity inlet with a negative direction
Exhaust air inlet	Pressure-outlet
Radiation heat	Discrete Ordinates (Do) radiation model

In the first ten cases, the supply air temperature is constant at 18℃ and the set-point for the air temperature at the head level is 25℃. The total supply airflow rate is estimated to be 2.75kg/s based on the assumption that 80% of the space cooling load is dispersed into the occupied zone. The fresh air required for the occupants is calculated according to Equation (6-2). R_p and R_a are set to 3.8L/s • person and $0.3L/s \cdot m^2$, respectively. The zone air distribution effectiveness E_z is set to 1.1 and the outside air flow rate is constant at 0.744kg/s. The humidity of the supply air is assumed to be kept constant at 50%.

The solar radiation is intensified in Cases 11 and 12. The heat gain from the right-hand glass wall is set to $150W/m^2$, considering the relevant solar radiative, free convective and conductive heat transfer from outside to inside that is absorbed by the glass wall with an absorbance of 0.18. The transmitted solar radiation into the space is 118W/m^2, with a glass wall transmissivity of 0.7. This portion of thermal load is considered by setting a fixed heat flux to the floor region exposed to the radiation, as indicated in Figure 7.3. Therefore, to maintain the same set-point indoor air temperature, an increase in the supply air volume of up to 3.96kg/s is required in these two cases. When the air is supplied from two different diffuser locations, such as in Cases 3 and 4, the velocity and flow rate of supply air for each diffuser are set to be the same in all cases.

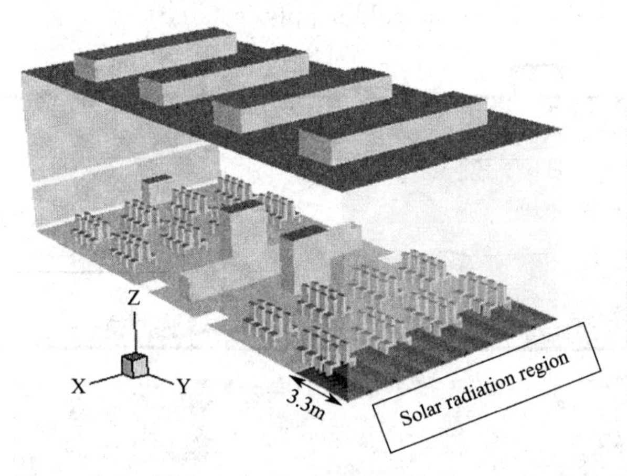

Figure 7.3 The transmitted solar radiation on the floor

7.3 Simulation results

7.3.1 Temperature and velocity distributions

7.3.1.1 Different air distribution designs

In the first design, with floor level air supply, the air flow in the occupied zone is dominated by the thermal buoyancy of the heat sources, while in the second design, with mid-height air supply (approximately 2m above floor level), the momentum flux from the supply diffusers is the main driving force of the air flow in the occupied zone. Figure 7.4 compares the temperature and velocity distributions for these two typical air distribution designs at Plane X=5.6m, which runs through the passageway and the air conditioning units. The shapes of occupants are also given in the figure for better perspective. The thermal stratification occurs in the entire space in Case 2, in which the air is supplied from Diffuser 1 and returned in Diffuser 4. The thermal stratification is especially distinct in the occupied zone, and the temperature difference between head level and ankle level was more than 3℃, which may result in the occupants' discomfort. In comparison, when the air is supplied from Diffuser 2 and returned in Diffuser 3 in Case 6, it mixes well in the occupied zone. The thermal environment in Case 6 is much more homogeneous than that in Case 2 and the thermal discomfort that occurs due to an excessively large temperature difference between head and ankle levels is avoided. But the air mixing in the occupied zone is not as complete as that in a small space and temperature variations exist horizontally. As shown in Figure 7.4 (*b*), the occupants seated close to the air conditioning units and below the supply jets may not obtain enough cooling and will experience a temperature that is too warm. Increasing the supply air velocity may enhance the air mixing and alleviate the local thermal island phenomenon. However, it is also possible to increase the local draft risk for the occupants exposed directly to the cold supply jets.

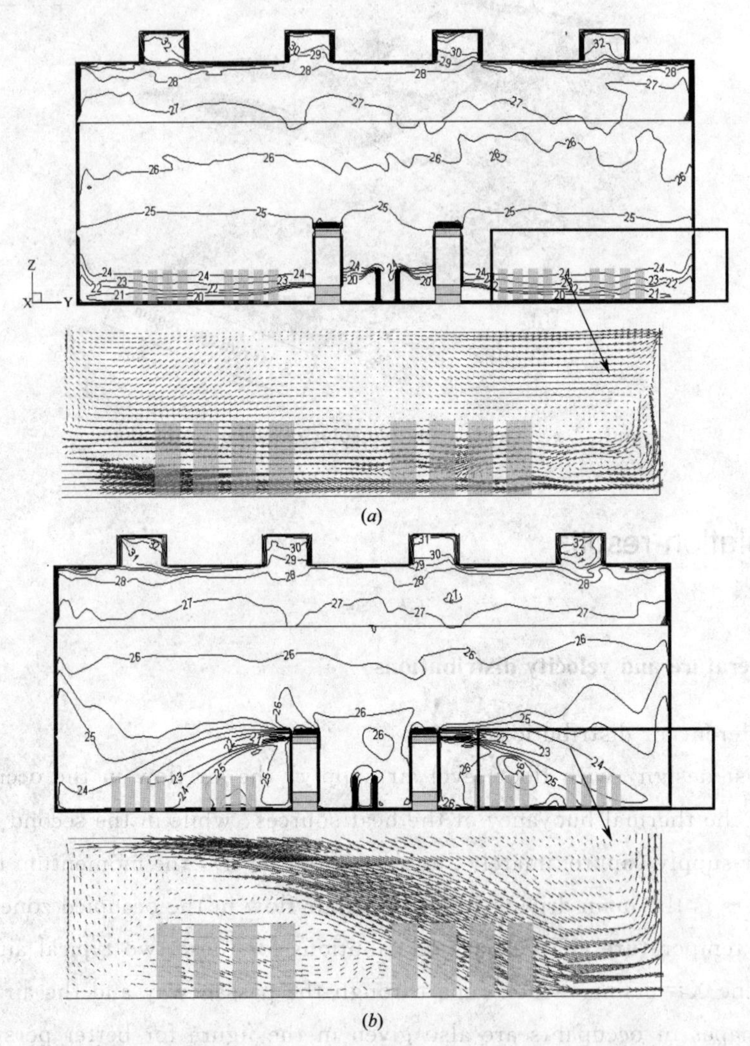

Figure 7.4 The temperature and velocity distributions at Plane X=5.6m in Case 2 and Case 6

(*a*) Case 2; (*b*) Case 6

7.3.1.2 Different supply diffuser locations

The locations of supply diffusers strongly affect the air flow fields in the space. Figure 7.5 presents results for the temperature and velocity distributions in Cases 1 and 3 at Plane X=5.6m. All of the air is supplied from Diffuser 1 in Case 1, while the air is supplied from Diffusers 1 and 3 simultaneously in Case 3. The air is returned in Diffuser 2 in both cases. The temperature gradient for the occupied zone in Case 3 is apparently smaller than that in Case 1, which is beneficial for occupant thermal comfort. In Case 1, the cold air is supplied only from Diffuser 1, located at the bottom of the air conditioning units. To avoid local drafts experienced by the occupants seated close to the air conditioning unit, the momentum flux of supply air is relatively small. Correspondingly, the air flow in the occupied zone in Case 1, which is dominated by thermal buoyance, is slight with less mixing, while in Case 3 the air is supplied from the bottom of the air conditioning units and the exterior walls simultaneously at a lower velocity than that in Case 1. When these two airstreams meet, reversed air flows are formed as shown in Figure 7.5 (*b*). The reversed air flows enhance the air mixing in the lower region of the occupied zone, and thus reduce the temperature difference between the head and ankle levels. Therefore, the temperature distribution in the occupied zone is more uniform.

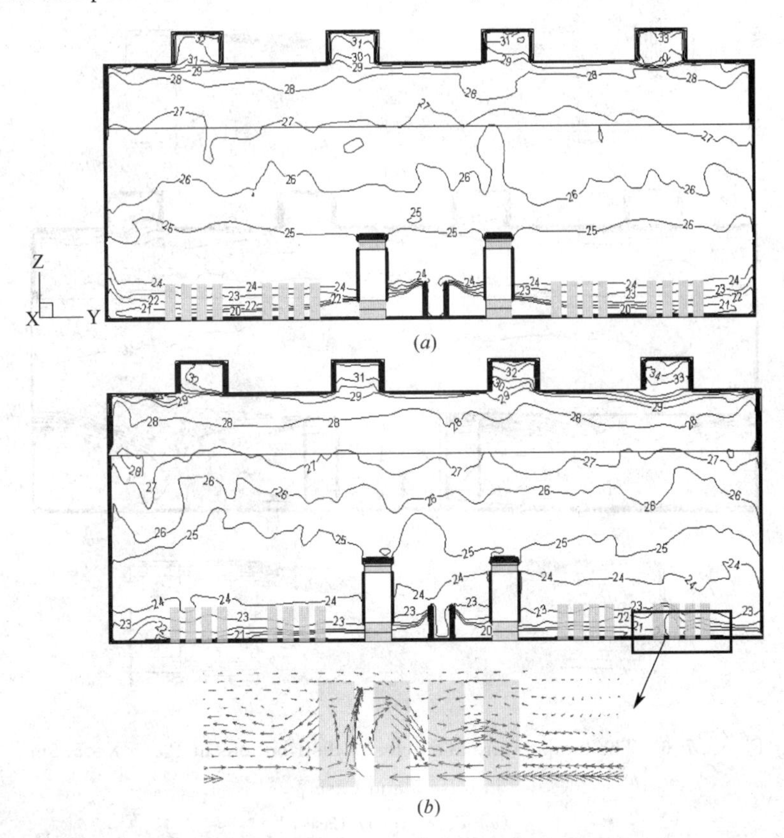

Figure 7.5 The temperature and velocity distributions at Plane X=5.6m in Case 1 and Case 3

(*a*) Case 1; (*b*) Case 3

Similar findings are also occur when the air is supplied at a height approximately 2m above floor level and returned at floor level, as shown in Figure 7.6. In Case 5, the air is supplied from Diffuser 2 alone, while in Case 7 the air is supplied from Diffusers 2 and 4 simultaneously. It is obvious that the air mixing in the occupied zone is much better in Case 7 due to the reversed air flow of the two supply airstreams. As a result, the temperature distribution in the occupied zone is more uniform. The thermal stratification in the upper zone is only slightly influenced by this scenario.

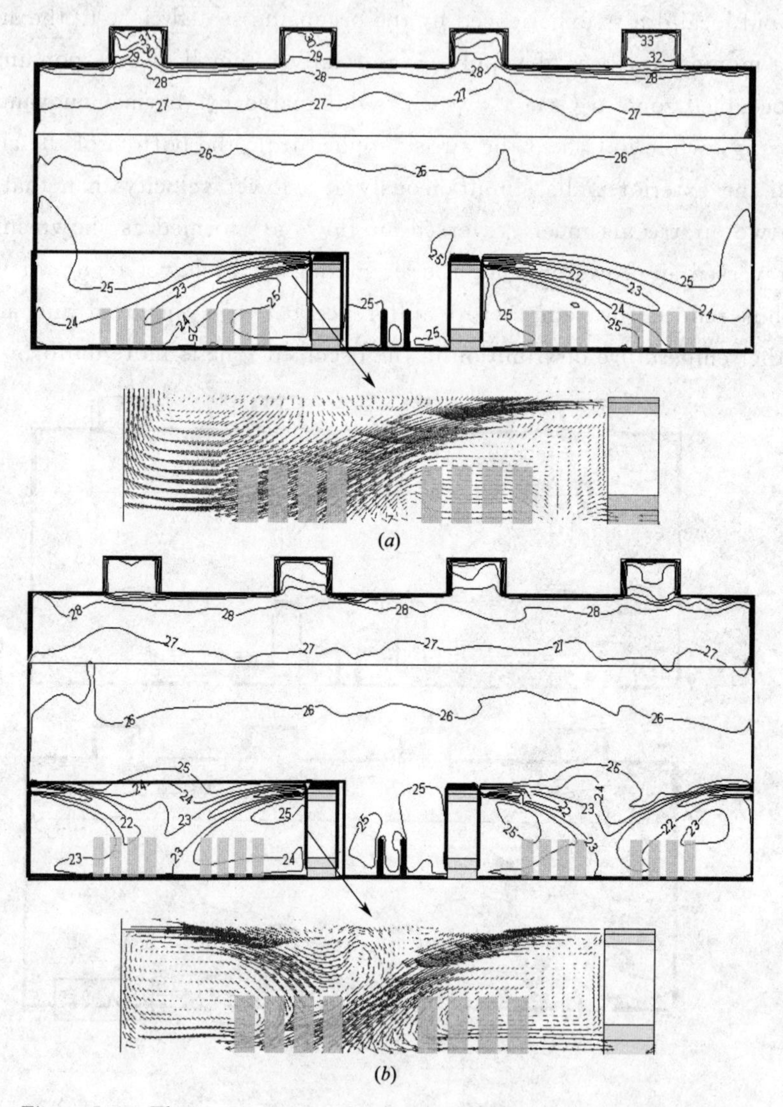

Figure 7.6 The temperature and velocity distributions at Plane X=5.6m in Case 5 and Case 7

(*a*) Case 5; (*b*) Case 7

7.3.1.3 Different return grille locations

Figure 7.7 presents the temperature distributions in Cases 4 and 9. In these two cases, the air is supplied at floor level from Diffusers 1 and 3 simultaneously, which is the same as the scenario in Case 3. The only difference is the return grille locations. The air is returned in Diffuser 4 in Case 4 and Diffusers 2 and 4 simultaneously in Case 9. Figure 7.7

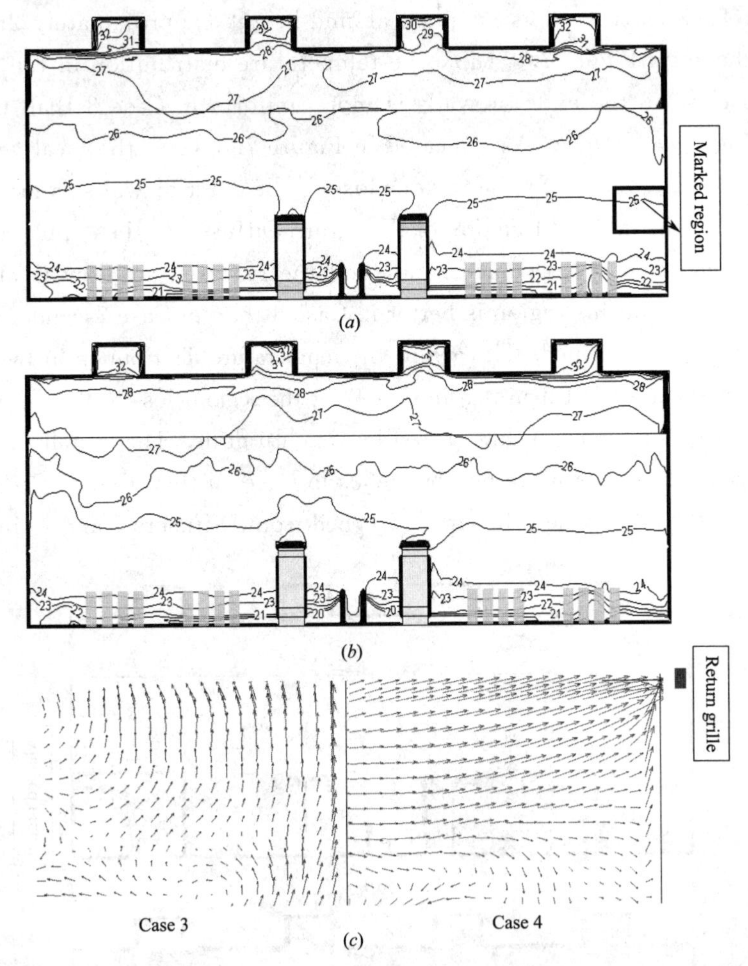

Figure 7.7 The air flow fields at Plane X=5.6m in Case 3, Case 4 and Case 9

(*a*) Case 4; (*b*) Case 9; (*c*) Local velocity vectors in Case 3 and Case 4

reveals that the thermal stratifications in Cases 4 and 9 are very similar. The temperature gradient in the occupied zone of Case 9 is slightly smaller than that in Case 4. However, the temperature distributions in the occupied zone in Cases 4 and 9 are much more stratified than that in Case 3, which is presented in Figure 7.5. The main reason for this difference is that the induced function of the return Diffuser 4 blocks the free convection flow of warm air along the exterior walls. Figure 7.7 (*c*) compares the local velocity vectors in Cases 3 and 4 for the region around the return grille. It can be observed that the cold air in Case 3 is warmed by heat sources and flows upward to the upper zone close to the exterior walls, while the induced function of the return Diffuser 4 in Case 4, located at the exterior walls, blocks the upward air flow and even leads a portion of air to flow backward and downward to the occupied zone, which increases the occupied zone's cooling load. Therefore, much more heat is transferred to the occupied zone in Case 4 than in Case 3 and results in an increase in head level air temperature and a larger temperature gradient.

In comparison, a positive effect on the thermal distributions is caused by returning air from Diffuser 4, when the air is supplied at mid-height (approximately 2m above floor level). As shown in Figure 7.8 (*a*), the temperature distribution in the occupied zone for the region close to the exterior walls is more uniform in Case 8 than that in Case 7 presented in Figure 7.6 (*b*). As indicated in Figure 7.8 (*c*), the local velocity vectors clearly vary in Cases 7 and 8 for the region close to the exterior walls as marked in Figure 7.8 (*a*). Due to the induced function of the return Diffuser 3, the supply air streamlines in Case 8 bend more to the occupied zone, and the reversed air flow is also enhanced. Thus, the air mixing in this region is better in Case 8 than in Case 7, and a more uniform thermal environment is formed. However, the temperature distribution in the occupied zone in Case 8 is more asymmetrical than that in Case 7 for the region close to the the air conditioning units, because there is no induced flow caused by a return grille. The overall thermal environment in the occupied zone is more homogeneous in Case 10 than that in Cases 7 and 8, as shown in Figure 7.8 (*b*), when the air is returned from Diffusers 1 and 3 simultaneously.

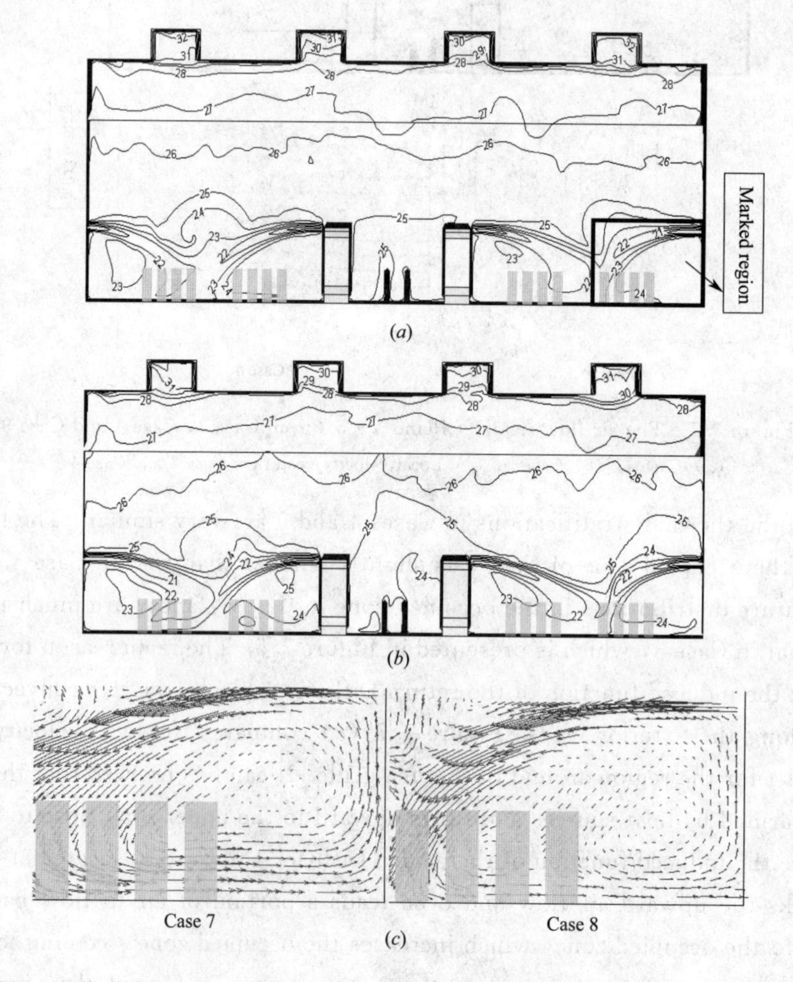

Figure 7.8 The air flow fields at Plane X=5.6m in Case 7, Case 8 and Case 10

(*a*) Case 8; (*b*) Case 10; (*c*) Local velocity vectors in Case 7 and Case 8

7.3.2 Thermal comfort

7.3.2.1 Different air distribution designs

Figure 7.9 compares the draft risk (DR) at Plane X=5.6m in Cases 2 and 6. The DR values are less than 20% for most of the occupied zone in Case 2, which meets the requirements listed in ASHRAE standard 55-2010. But a potential risk of local thermal discomfort caused by draft does exist in the regions near the supply diffusers located at the bottom of air conditioning units. However, the draft risk can be alleviated because no occupants are directly exposed to the cold supply air, as indicated from the predicted percentage of dissatisfied (PPD) at Plane Z=0.6m presented in Figure 7.10 (*a*). In Case 6, the local draft risk for the occupants exposed directly to the cold supply airstreams, is manifested in both Figure 7.9 (*b*) and Figures 7.10 (*b*). The DR values in this case are as high as 30, greatly exceeding the limited value.

7.3.2.2 Different supply diffuser locations

The influence of supply diffuser location on the velocity and temperature distribution

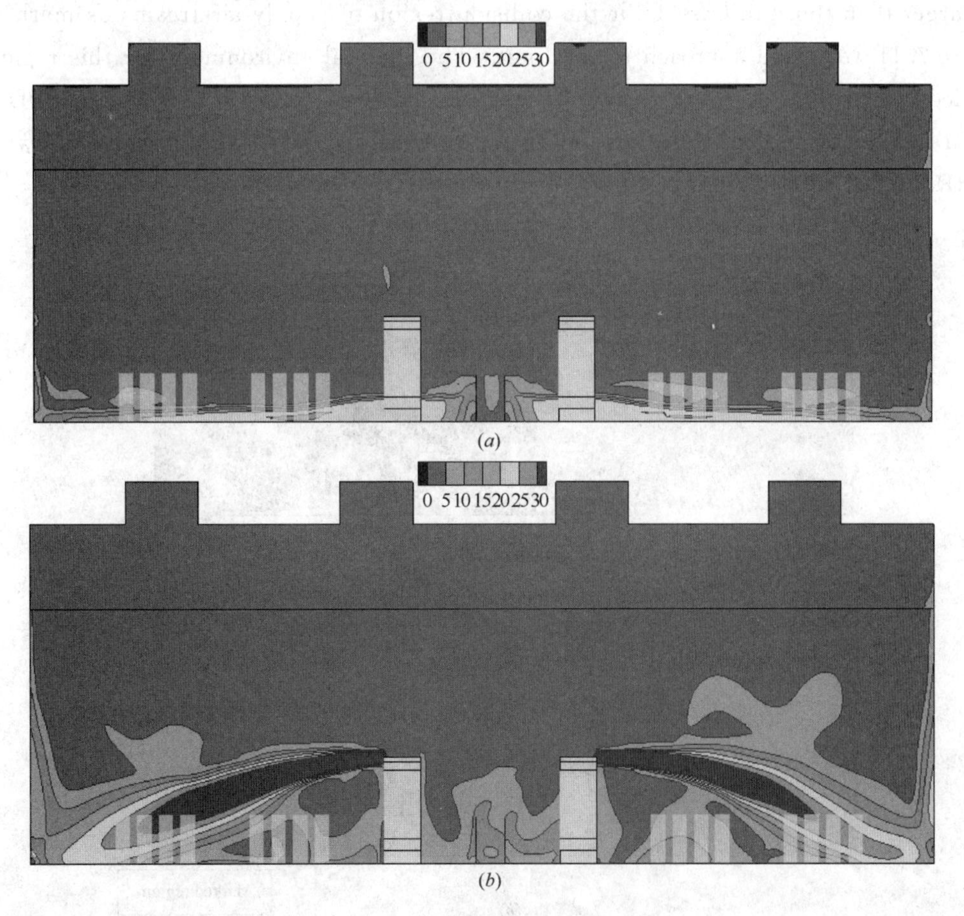

Figure 7.9 DR values at Plane X=5.6m in Case 2 and Case 6

(*a*) Case 2; (*b*) Case 6

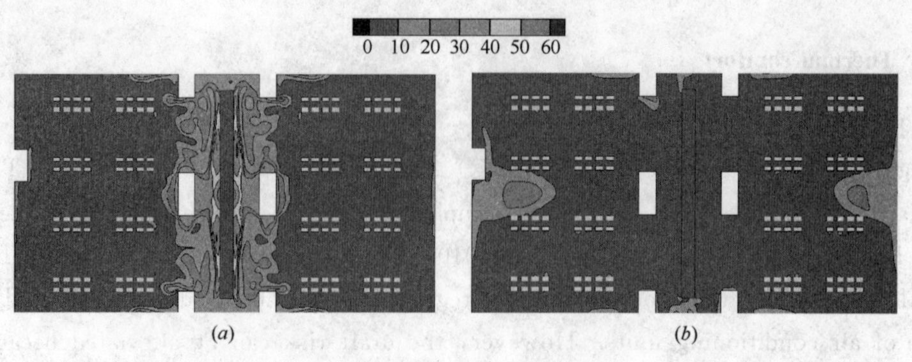

Figure 7.10 PPD values at Plane Z=0.6m in Case 2 and Case 6

(a) Case 2; (b) Case 6

of flow fields is studied in Section 7.3.1.2, when the air is supplied from the bottom of the air conditioning units. It was found that locating additional supply diffusers at exterior walls alleviates the thermal stratification in the occupied zone, because the collision of the two supply airstreams strengthens the disturbance of the flow field. However, the enhanced turbulence intensity also increases the local draft risk. The DR values in Case 3 are larger than those in Case 1 for the collision region of supply airstreams as marked in Figure 7.11 (b), and approach to the limit. The thermal environment for this region is verified to be poorer in Case 3 than in Case 1. As shown in Figure 7.12 (b), the PPD values in this region in Case 3 are greater than 10, which is the maximum value allowed in ASHRAE Standard [2010].

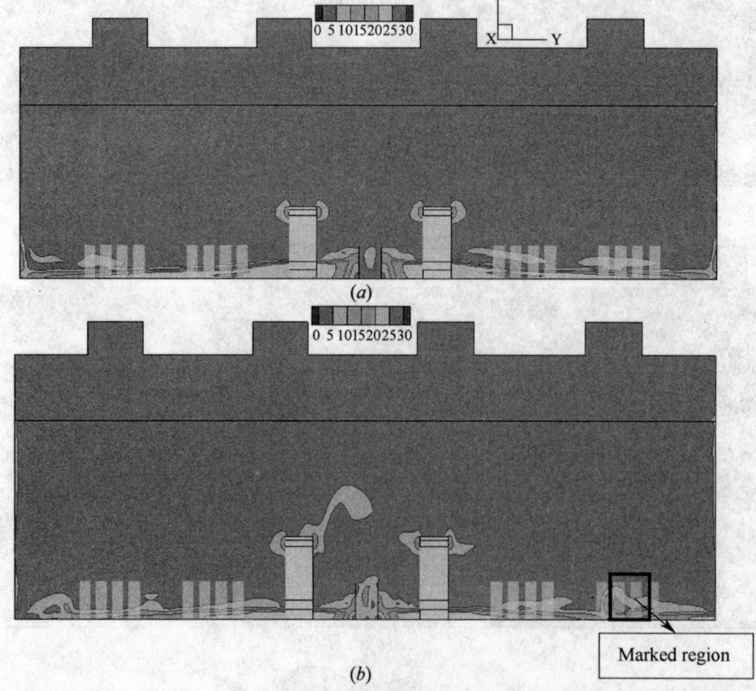

Figure 7.11 DR values at Plane X=5.6m in Case 1 and Case 3

(a) Case 1; (b) Case 3

It is interesting that the thermal environment satisfies the requirements for symmetrical region shown in Figure 7.12 (*b*). Due to the entrance and ticket counter layouts, supply diffusers located at the exterior wall and air conditioning unit are staggered on the left side, thus avoiding the direct collision of the two supply airstreams, which helps reduce the local draft risk. However, the local draft is largely alleviated in Case 3 compared to Case 1 for the region close to the air condition units, as expected, because by locating additional supply diffusers at the exterior walls in Case 3, the momentum flux of the supply air from the air conditioning units is only approximately half that of Case 1.

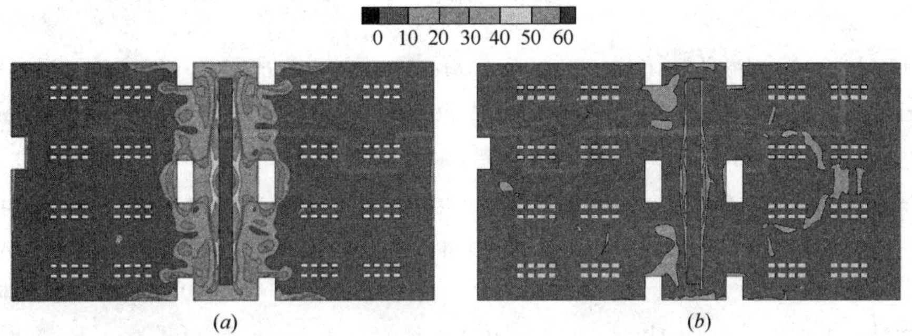

Figure 7.12 PPD values at Plane Z=0.6m in Case 1 and Case 3

(*a*) Case 1; (*b*) Case 3

Similar findings have also been made when the air is supplied at mid-height as shown in Figure 7.13. By locating additional supply diffusers at exterior walls in Case 7, the air disturbance in the occupied zone is enhanced and better air mixing is achieved than in Case 5, when supplying air from Diffuser 2 alone. However, the local draft area is expanded due to the collision of the two supply airstreams. Consequently, there is a possibility that some occupants may complain about the local thermal environment, as indicated in Figure 7.14 (*b*).

7.3.2.3 Different return grille locations

Figure 7.15 presents the DR value distributions at Plane X=5.6m in Cases 4 and 9. As indicated in Figure 7.15, the local draft risk region caused by the direct collision of the two supply airstreams also exists in Cases 4 and 9. However, this region moves slightly to the right side and closer to the exterior wall in Cases 4 and 9 compared to that in Case 3,

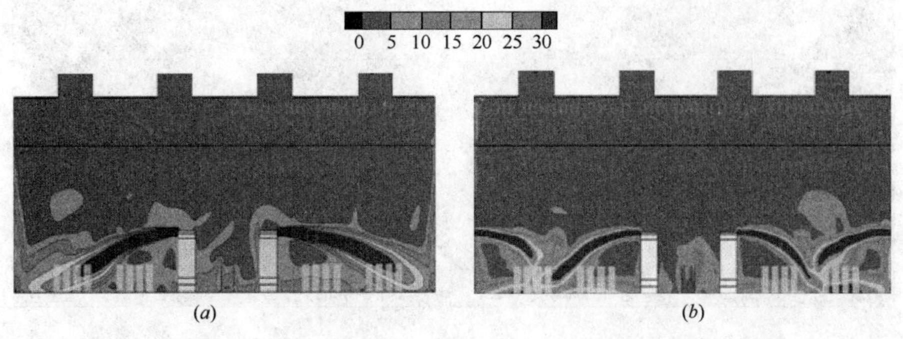

Figure 7.13 DR values at Plane X=5.6m in Case 5 and Case 7

(*a*) Case 5; (*b*) Case 7

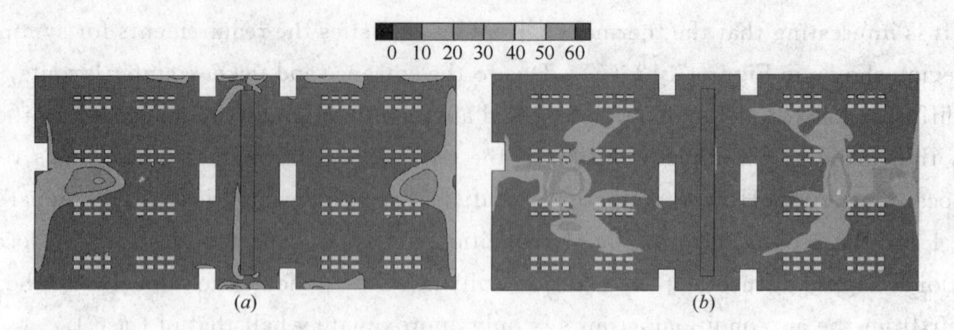

Figure 7. 14 PPD values at Plane Z=0. 6m in Case 5 and Case 7

(a) Case 5; (b) Case 7

presumably because the flow field areas that are directly affected by the induced function of the return grilles and the collision areas of supply airstreams move toward the exterior walls. Accordingly, the thermal regions of discomfort are also shifted slightly to the exterior walls in Cases 4 and 9 as indicated in Figure 7. 16. Thus, it can be concluded that when air is supplied at floor level and returned at mid-height in a large space with high ceilings, as adopted in Cases 3, 4 and 9, the return grille locations have little impact on the thermal environment of the occupied zone.

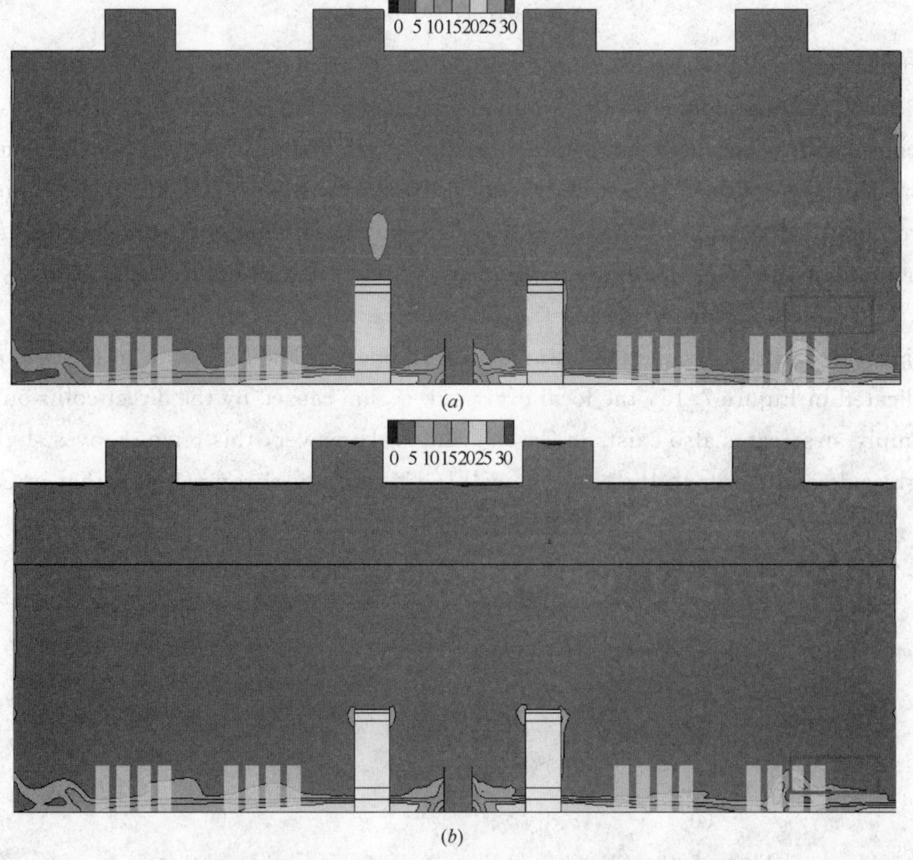

Figure 7. 15 DR values at Plane X=5. 6m in Case 4 and Case 9

(a) Case 4; (b) Case 9

Figure 7.16 PPD values at Plane Z=0.6m in Case 4 and Case 9

(*a*) Case 4; (*b*) Case 9

Figure 7.17 presents the DR value distributions at Plane X=5.6m in Cases 8 and 10 when the air is supplied from Diffusers 2 and 4 simultaneously and returned at floor level. Comparing the DR value distributions in Case 8 with those in Case 7, which are presented in Figure 7.13 (*b*), it is found that the induced function of the return grilles alleviates the local draft risk to a large extent, which is especially obvious on the right side of the occupied zone. Because the location of return grille is switched from Diffuser 1 in Case 7 to Diffuser 3 in Case 8, the local draft regions move correspondingly from the areas

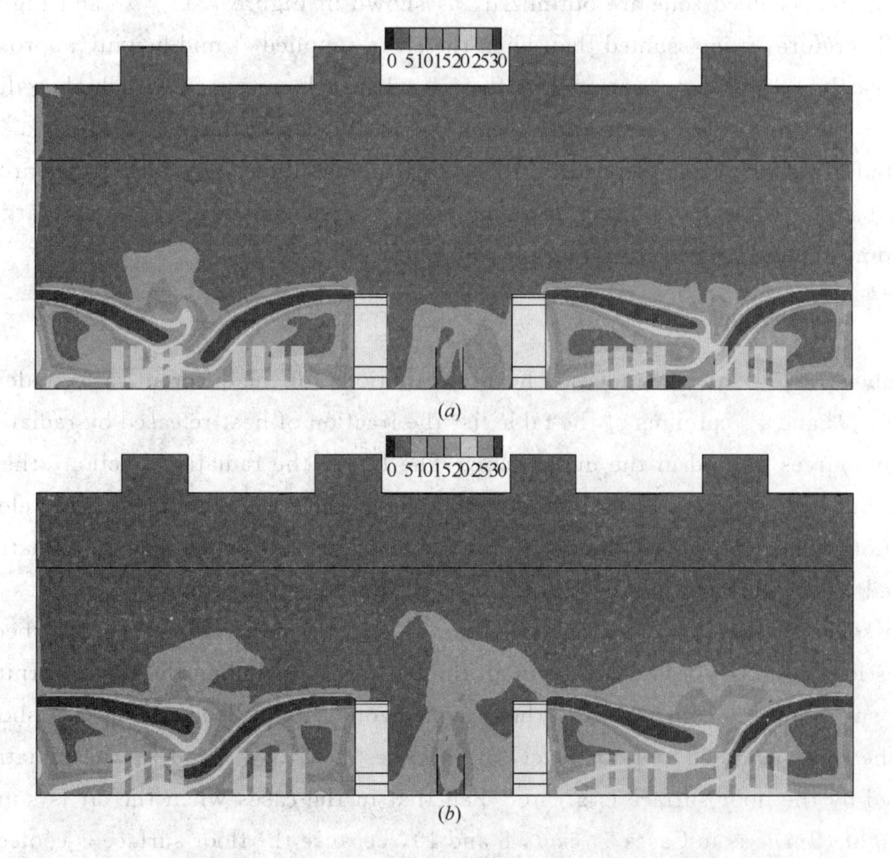

Figure 7.17 DR values at Plane X=5.6m in Case 8 and Case 10

(*a*) Case 8; (*b*) Case 10

close to the exterior walls to the areas near the air conditioning units. In Case 8, the thermal environment of the regions close to the exterior walls is greatly improved as a result, but largely worsened in the regions near the air conditioning units, compared with that in Case 7.

Figure 7.18 PPD values at Plane Z=0.6m in Case 8 and Case 10

(*a*) Case 8; (*b*) Case 10

This effect is also demonstrated by comparing the PPD value distributions at Plane Z=0.6m in Figure 7.14 (*b*) for Case 7 with those in Figure 7.18 (*a*) for Case 8. In Case 10, when the air is simultaneously returned from Diffusers 1 and 3, the thermal environments of the occupied zone are optimized, as shown in Figure 7.17 (*b*) and Figure 7.18 (*b*). Therefore, it is assumed that when the air is supplied at mid-height (approximately 2m above floor level) and returned at floor level in a large space with high ceiling, the induced function of the return grille is beneficial to alleviating the local draft risk. Thus, a more uniform distribution of return grilles, such as installing the return grilles around the occupied zone with larger number and smaller area versus smaller number and larger area, is recommended to improve the thermal environment.

7.3.3 Energy savings

Table 7.4 presents the radiation heat distributions for the different cases under study. The 2nd, 3rd and 4th columns of the table list the fraction of heat released by radiation from the heat sources located in the upper zone. The larger the radiative fraction, the greater the amount of heat released via radiation and the smaller the amount of heat released by convection. The last two columns of the table show the percentage of radiation heat absorbed by the surfaces in the occupied zone compared to the radiation heat released from all heat sources. It is obvious that more than 95% of the radiation heat is absorbed by the surfaces located in the occupied zone in all the ten cases, including the floor, furniture and facility surfaces, and contributes to the occupied zone cooling load. Table 7.4 shows that when the air is supplied at floor level, as in Cases 1, 2, 3, 4 and 9, the radiation heat absorbed by the floor surface is greater than that in the cases when the air is supplied at mid-height (2m), as in Cases 5, 6, 7, 8 and 10, because the floor surface is cooled directly by the cold supply air in the cases where the air was supplied at floor level, and thus the surface temperature is lower than that in the cases when the air is supplied at mid-height

(2m). Table 7.4 also verifies the findings indicating that the free convection flow along the exterior wall is impaired by locating additional return grilles at the exterior walls when supplying air at floor level, which is demonstrated in Section 7.3.1.3. For instance, the radiative fraction for the exterior wall located at the upper zone is 23.5% in Case 3, while the value increases dramatically to 48.1% in Case 4. Therefore, the convective fraction for the exterior wall located at the upper zone is proportionally reduced, which indicates a weakening of the free convection flow along the exterior wall.

Radiation heat distributions in different cases **Table 7.4**

	Radiative fraction of heat sources located at upper zone			Radiative heat gain in the occupied zone	
	Ceiling &Skylight	Lightings	Side wall (Upper zone)	Floor	Furniture & Facility surfaces
$Q_{i-space}$ (W)	4113	4390	4114		
Case 1	72.1%	80.4%	30.1%	81.3%	15.3%
Case 2	68.7%	79.7%	42.7%	82.3%	15.0%
Case 3	66.0%	78.3%	23.5%	84.5%	13.2%
Case 4	66.3%	79.0%	48.1%	80.6%	17.0%
Case 5	74.7%	80.1%	25.9%	76.0%	21.1%
Case 6	74.0%	79.7%	27.7%	76.3%	20.6%
Case 7	72.4%	79.3%	34.9%	77.3%	19.5%
Case 8	76.0%	81.7%	26.9%	79.1%	18.4%
Case 9	66.9%	79.0%	46.0%	80.9%	16.8%
Case 10	73.0%	80.6%	26.9%	77.3%	19.5%

Because only the lower occupied zone must be cooled in the terminal building, while the un-cooled upper zone accounts for the greater fraction of the space, the energy savings potential for the building is considerable with a STRAD system. When the air is supplied at floor level, as in common stratified air conditioning systems, the energy savings of the cooling coil can be evaluated directly by using Equation (4-9). When the air is supplied at mid-height (2m), Equation (4-3) to Equation (4-7) are still applicable. Thus, the energy savings in such conditions can be estimated according to Equation (4-9), by calculating the cooling coil load reduction ΔQ_{coil}. The cooling coil load reductions for the different cases are presented in Table 7.5. It appears that locating additional supply diffusers at the exterior walls increases the cooling coil load reduction ΔQ_{coil} when the cold air is supplied at the floor level. For instance, the reduced cooling coil load increases from 16.5% of the space cooling load in Case 2 to 17.2% in Case 4. A similar conclusion can also be drawn for the systems that supply air at mid-height, approximately 2m above the floor level. For example, the reduced cooling coil load is 15.4% of the space cooling load in Case 5, while the value increases to 16.6% in Case 7.

Energy-savings for cooling coil Table 7.5

Cases	t_r(℃)	t_e(℃)	ΔQ_{coil}(W)	$\Delta Q_{coil}/Q_{Space}$
Case 1	24.9	30.5	4094	16.8%
Case 2	25	30.4	4020	16.5%
Case 3	24.66	31.8	5047	20.8%
Case 4	24.9	30.6	4176	17.2%
Case 5	25.1	30	3737	15.4%
Case 6	25	30.2	3886	16.0%
Case 7	24.8	30.4	4154	16.6%
Case 8	24.7	30.6	4042	17.1%
Case 9	24.9	30.7	4273	17.6%
Case 10	24.7	30.6	4199	17.3%
Case 11	25.73	34.1	6774	19.3%
Case 12	25.39	34.2	6871	19.5%

However, installing the return grille at the exterior walls impairs the energy-saving capacity of the ventilation systems, when the air is supplied at floor level. As revealed in Section 7.3.1.3, the induced effect of the return grilles located at the exterior walls blocks the upward flow of warm air along with the exterior wall. Therefore, much more heat is released to the occupied zone and adds to the cooling coil load. It is obvious that the cooling coil load reduction is reduced from 20.8% of the space cooling load in Case 3 to 17.2% in Case 4 as the air returning locations vary from Diffuser 2 to Diffuser 4. In comparison, locating the return grille at the exterior wall can enhance the energy efficiency of the ventilation systems when the air is supplied at mid-height (2m), but the effect is minimal. For instance, the reduced cooling coil load is 16.6% of the space cooling load in Case 7, but increases to 17.1% in Case 8. In the first ten cases, the energy savings of the cooling coil is most prominent in Case 3, in which it makes up as much as 20.8% of the space cooling load.

Two additional cases, 11 and 12, are analyzed to study the influence of different solar radiation intensity on the energy performance of the ventilation systems. In Case 11 the ventilation system is designed identically to that of Case 9, while in Case 12 the ventilation system is identical to that in Case 10. The space cooling load Q_{Space} for Cases 11 and 12 is calculated according to the raised solar radiant intensity in Table 7.5. The heat gains for the right-hand wall in Cases 11 and 12 reaches 150W/m^2 and the heat gains for the floor region exposed directly to the solar radiation, as indicated in Figure 7.3, are 118 W/m^2. Table 7.5 shows that, as the cooling load increases, the relative energy saving potential of the cooling coil $\Delta Q_{coil}/Q_{Space}$ increases for both typical air distribution designs. One of the main reasons is that as the solar radiation increases, the proportion of heat gain for the right-hand glass wall compared to the entire space cooling load also increases. The value is 10.2% in Cases 9 and 10, and increases up to 36.6% in Cases 11 and 12. The thermal buoyancy flow along the exterior wall is correspondingly strengthened. As indicated in Figure 7.19, much more warm air flows upward along the exterior wall to the un-occupied zone in Case 11 than in Case 9, which indicates that much more heat is discharged from the occupied zone and excluded from the cooling coil load. The location of the exhaust air grilles at the region close to the skylight is essential as more heat accumulates in the upper zone.

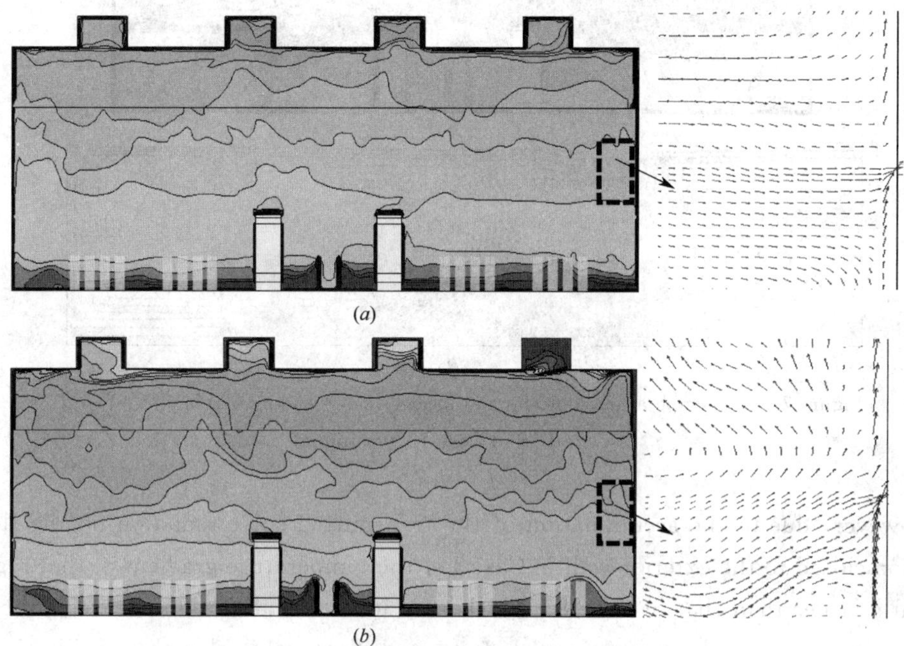

Figure 7.19 Temperature contours and local velocity vectors at Plane X=5.6m in Case 9 and Case 11

(*a*) Case 9; (*b*) Case 11

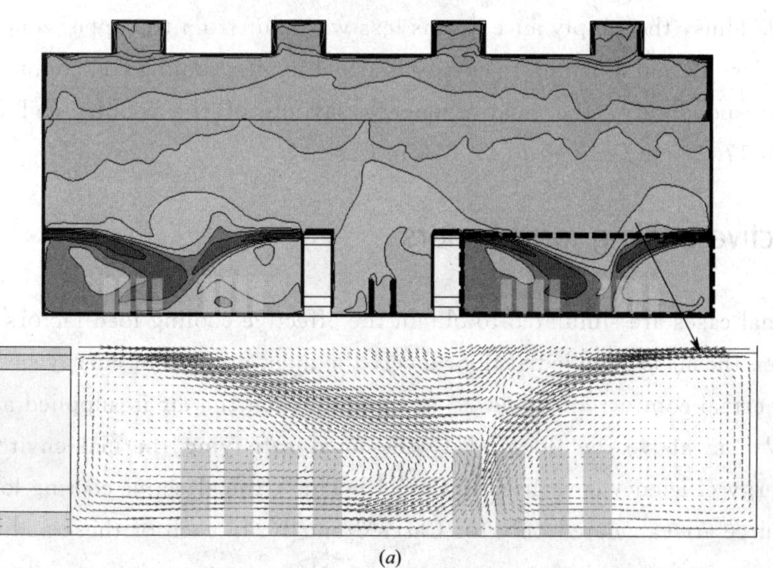

Figure 7.20 Temperature contours and local velocity vectors at Plane X=5.6m in Case 10 and Case 12

(*a*) Case 10

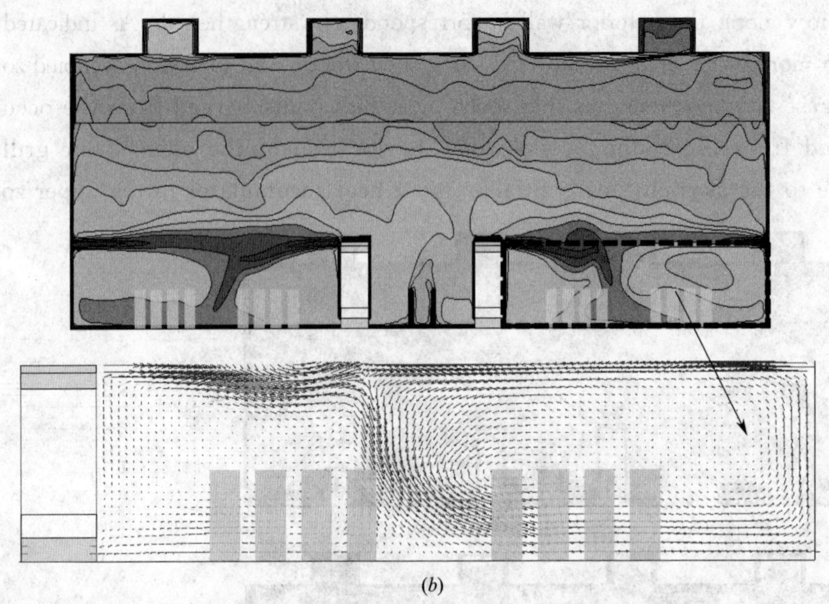

(*b*)

Figure 7. 20 Temperature contours and local velocity vectors at Plane X=5. 6m in Case 10 and Case 12 (continue)

(*b*) Case 12

However, due to the accumulation of the transmitted solar radiation on the floor in the region closed to the exterior wall in Case 11, the temperature gradient in that region is apparently larger than that in Case 9, which may lead to undesirable thermal problems. The intensified thermal buoyancy flow along the exterior wall is also observed in Case 12. As shown in Figure 7. 20, the streamlines of supply air are much more linear in Case 12 than in Case 10 because the thermal plumes flowing from the floor level to the upper zone are enhanced. Thus, the supply air entrains less warm air from the upper zone to the occupied zone in Case 12 and much more energy is saved for the cooling coil compared to that in Case 10. Correspondingly, the relative energy savings of the cooling coil $\Delta Q_{\text{coil}}/Q_{\text{Space}}$ increase from 17. 3% in Case 10 to 19. 5% in Case 12.

7. 4 Effective cooling load factors

Additional cases are simulated to obtain the effective cooling load factors of each heat sources. When the air is supplied at floor level, the most prominent energy savings without sacrificing thermal comfort are achieved in Case 3. When the air is supplied at mid-height approximately 2m above the floor level, the most satisfying thermal environment with excellent energy efficiency is realized in Case 10. Thus, the effective cooling load factors of each heat source are calculated only for Cases 3 and 10. In each of these additional cases, there is only one maintained heat source and the others are turned off, while the CFD settings are identical to those in Cases 3 or 10. The simulation results are presented in Table 7. 6.

Effective cooling load factors of each heat sources for Case 3 and Case 10　　Table 7.6

Heat Gains	Q_i(W)	$ECLF_i$ ($Q_{i\text{-occupied}}/Q_{i\text{-speace}}$)	
		Case 3-based	Case 10-based
Ceiling & Skylights	4113	0.66	0.67
Horizontal elevator	800	0.86	0.80
Facilities	800	0.83	0.70
Lightings	4390	0.71	0.68
Occupants	8960	0.84	0.88
Exterior walls	4757	0.63	0.62
$Q'_{occupied}$(W)		17686	17759
$Q_{occupied}$(W)		17272	19460
$(Q'_{occupied}-Q_{occupied})/Q_{occupied}$		1.7%	−7.2%

It is found that the effective cooling load factors of most of the heat sources in the two configurations in Cases 3 and 10 are very close. When the air is supplied at floor level, the effective cooling load factor of the facilities is slightly larger than that in which the air is supplied at mid-height approximately 2m above floor level. When the air flow in the occupied zone is mainly driven by thermal buoyancy, much more of the heat released from the facilities is detained in the occupied zone and contributes to the occupied zone cooling load compared to that in the cases where the air flow in the occupied zone is dominated by the momentum flux of supply diffusers.

The occupied zone cooling loads $Q_{occupied}$ are calculated according to Equation (5-1), while the effective occupied zone cooling loads $Q'_{occupied}$ are calculated according to Equation (2-12). For both typical air distribution designs, the divergences between $Q_{occupied}$ and $Q'_{occupied}$ are small enough to meet the engineering requirement, which re-confirms that the occupied zone cooling load calculation method proposed by Xu et al. [2009] is suitable for the application of STRAD systems in spaces with large floor areas and high ceilings. The occupied zone cooling load in Case 10 is larger than that in Case 3 because the supply air jets in Case 10 flowing from mid-height (approximately 2 m above floor level) contain high momentum flux and restrain thermal plume flow generated from heat sources located in the occupied zone. Furthermore, the supply air jets in Case 10 also entrain more warm air from the upper zone to the occupied zone and contribute to the occupied zone cooling load compared to that in Case 3.

7.5 Conclusions

This chapter studied the thermal environment in a hypothetical terminal building with a large floor area and high ceiling. Two typical air distribution designs, the supplying of air at floor level or at mid-height (approximately 2m above the floor level), were investigated in terms of thermal comfort and energy savings. The impact of the diffuser locations and solar radiation on the ventilation performance was illustrated.

When the air was supplied at floor level and returned at mid-height (approximately 2m above floor level), a distinct thermal stratification was identified in the occupied zone and thermal discomfort may have been caused by an excessively large temperature gradient. A more uniform distribution of supply air diffusers, such as by installing a larger number of supply diffusers with smaller areas around the occupied zone, rather than a smaller number with larger areas, could reduce the temperature gradient. The temperature gradient in the occupied zone was smaller when locating additional supply diffusers at the bottom of exterior walls compared to supplying air from the bottom of the air conditioning unit alone. The energy efficiency of the ventilation system was only slightly influenced by a more uniform distribution of supply air diffusers, but special attention should be given to the airstream collision regions to avoid un-comfortable local drafts. The locations of return grilles had no obvious effects on the thermal environments of the tested conditions. Whereas, locating the return grilles at exterior walls blocked the free upward convection of warmed air along the exterior walls, and much more heat released to the occupied zone, increasing the cooling coil load. When the air was supplied at mid-height (approximately 2m above the floor) and returned at floor level, the thermal environment was more uniform in the occupied zone compared to that when the air was supplied at floor level and returned at mid-height. But a high local draft risk existed for the occupants exposed directly to the cold supply air jets.

As the solar radiation intensity was raised, the upward thermal buoyancy flow along the exterior wall was strengthened, which was beneficial to the cooling coil load reduction. The intensified thermal buoyancy flow in the occupied zone uncurled the streamlines of supply air when the air was supplied at mid-height (approximately 2m above the floor level), which was beneficial to reducing the entrainment of warm air from the upper zone to the occupied zone. Therefore, the energy efficiency of the system was further improved. However, the temperature gradient in the region exposed directly to the transmitted solar radiation also increased, when the air was supplied at floor level. Thus, external shading designs are particularly important in this configuration.

Chapter 8 Conclusion and recommendations for future work

8.1 Conclusions

In this book the thermal and airflow processes of stratified air distribution (STRAD) systems are experimentally and numerically illustrated. The applications of STRAD systems with separate locations of return and exhaust grilles in three different spaces, namely office space with low ceiling level (lower than 3m), high space with middle ceiling height (5m) and large space with very high ceiling level (8m), were numerically investigated with regard to both thermal comfort and energy saving. Several crucial design parameters for STRAD systems were studied, including the locations of diffusers, the occupied zone cooling load that will be used to determine the required supply air flow rate, and the cooling coil load that will be used to determine the equipment size, and significant results were obtained. As concluding remarks are already given at the end of each chapter, only the main points are highlighted here in this section.

1) The thermal environment in a small office ventilated by a STRAD system was assessed by coupling CFD simulations with two representative thermal comfort models, namely the simple ISO standard 14505 method and the comprehensive UCB thermal comfort model. The evaluation results were compared. The comparison results indicated that in an overall thermally neutral environment, the ISO Standard 14505 method was more sensitive to the warming variation, but less sensitive to the cooling variation, in comparison with the UC Berkeley thermal comfort model. As a convenient way, the ISO standard 14505 method does not include the thermoregulation model and the reliability of the method should be treated with caution when evaporation from skin is involved. In comparison, the coupling procedure between the comprehensive UC Berkeley thermal comfort model and CFD simulation is complicated, and an iteration process is required. Thus, an innovative coupling procedure is proposed with simplified operation and improved evaluation accuracy.

2) The energy saving principle for STRAD systems is clearly illustrated. The energy saving of the cooling coil is associated with but different from the space cooling load reduction. By splitting the locations of return and exhaust grilles, further energy saving of the cooling coil is available because much more heat can be discharged directly from exhaust grilles. The energy-saving potential is directly proportional to the exhaust air temperature. A novel CFD-based cooling coil load calculation method was developed, which could

be used to evaluate the cooling coil load reduction and determine the equipment size in STRAD systems. For the first time, experiments were conducted in a full-scale chamber ventilated by a STRAD system with separate locations of return and exhaust grilles. The experimental results demonstrated that the newly developed cooling coil load calculation method was feasible to be used in STRAD systems. The results also manifested that further energy was saved for the cooling coil by splitting the locations of return and exhaust grilles. In the experiments, as the location of return grille varied from ceiling level (2.4m) to middle level (1.85m), the energy saving of the cooling coil improved obviously, while the temperature difference between the head and ankle levels $\Delta t_{head\text{-}ankle}$ increased slightly.

3) CFD model validation procedures were carried out by using the experimental results as well as the data published from a previous literature. The results confirmed that the adopted CFD model was capable to evaluate the ventilation performance of STRAD systems. Based on that, numerical simulations were conducted to study the thermal environment in a hypothetical office located at the perimeter zone and conditioned by a STRAD system with separate locations of return and exhaust grilles. The influence of outdoor climate to the indoor air flow characteristic was considered. The simulation results manifested that the newly developed cooling coil load calculation method was feasible to be used when designing a SATRD system. The results also clearly demonstrated that maintaining the exhaust grille at ceiling level and locating the return grille at middle level achieved extra energy saving for a STRAD system. The lower the return grilles height, the higher the exhaust air temperature and the more the energy saved for the cooling coil. However, locating the return grille at the lower occupied zone and too close to the supply diffusers may lead to "short-circuit" of cold supply air and have a strongly negative effect on the thermal environment. Thus, it was recommended that the return grille be designed at the upper boundary of the occupied zone. The exhaust grilles should be located at ceiling level and close to the lighting fixture and exterior wall, which may enhance the upward free convective flow along the exterior wall and make much more heat to be discharged directly before entering into the occupied zone.

4) Different designs to realize thermal stratification in a large-space lecture theatre with terraced floor were numerically investigated. The simulation results indicated that for the head levels, occupants seated at back rows were exposed to much warmer air compared to that at front rows because of the thermal air plumes flowing from front to back rows along terraced seats. While for the ankle level, occupants seated at back rows tended to suffer from local cold draft because portion of the cold air supplied at floor level with low-velocity flowed downwards along with the terraced floor. Supplying air from desk-edge-mounted grilles achieved better thermal sensations for the lower body parts of the occupants compared with supplying air at floor level only, and thus the overall thermal comfort of the occupants was improved. Thermal environment was optimized when the air was sup-

plied from the three different locations simultaneously. Satisfied thermal environment could be maintained when locating the return grilles at the lower occupied zone. But it is notable that the return grilles should not be located too close to the occupants and worsen the thermal comfort of occupants. As the return grilles height decreased, the exhaust air temperature increased obviously and the energy saving of the cooling coil improved significantly. However, the temperature of reversed air flowing from the upper zone to the occupied zone also increased and the occupied zone cooling load $Q_{occupied}$ increased correspondingly.

5) The thermal environment in a hypothetical terminal building with large floor area and high ceiling was studied by using CFD simulations. Alternative ventilation designs, namely supplying air at floor level or at mid-height (approximately 2m above floor level), were investigated in terms of thermal comfort and energy efficiency.

When the air was supplied at floor level and returned at mid-height, distinct thermal stratification was realized in the occupied zone and thermal discomfort may be caused by an excessive large temperature gradient. A more uniform distribution of supply air diffusers, i. e. installing a larger number of supply diffusers with a smaller area each around the occupied zone versus a smaller number with larger area each, was able to reduce the temperature gradient in the occupied zone as well as to promote the energy efficiency of the ventilation system. But special attentions need to be given to the airstreams collision regions to avoid un-comfortable local draft. The locations of return grilles slightly influenced the thermal environments of the tested conditions. However, locating return grilles at the exterior walls blocked upward free convection of warmed air flowing along with the exterior wall from the occupied zone to the upper zone. Thus, much more heat was released to the occupied zone and contributed to the cooling coil load.

When the air was supplied at mid-height (approximately 2m above the floor level) and returned at floor level, the thermal environment was more uniform in the occupied zone, compared with that when the air was supplied at floor level and returned at mid-height. But a high local draft risk existed for the occupants exposed directly to the cold supply air jets flow.

Intensified solar radiation intensity was beneficial to improve the energy efficiency of STRAD systems. However, the temperature gradient in the region that exposed directly to the transmitted solar radiation increased when the air was supplied at floor level. Thus, a decent external shading design is much more important in such condition.

Based on the findings in this book, consultant engineers are suggested to take in account of the following special considerations when designing a STRAD system.

• Occupied zone cooling load calculation.

When determining the supply air flow rate in a STRAD system, only the heat gains contribute to the lower occupied zone and affect the thermal comfort of the occupants are need to be considered. The occupied zone cooling load calculation method proposed by Xu et al. [2009] is promoted and completed. The effective cooling load factors for different heat sources in three common types of spaces are summarized in Table 8.1, Table 8.2 and Table 8.3 respectively.

Effective cooling load factors in small office environments **Table 8.1**

Small office (≤3m)	Exterior wall	Windows	Heat source	Computer	Occupants	Lamps
$ECLF_i$	0.74～0.77	0.80～0.83	0.81～0.87	0.88～0.90	0.89～0.96	0.64～0.67
Note	The lower the return grilles locations, the lower the values used; The higher the return grilles locations, the higher the value used.					

Effective cooling load factors in large space with terraced floor **Table 8.2**

Large space with terraced floor (≤5m)	Occupants	Lightings
$ECLF_i$	0.83～0.90	0.70～0.82
Note	The lower the return grilles locations, the higher the values used; The higher the return grilles locations, the lower the value used.	

Effective cooling load factors in large space with high ceiling **Table 8.3**

Space with large floor area and high ceiling (>5m)	Ceiling & Skylights	Horizontal elevator	Facilities	Lightings	Occupants	Exterior walls
$ECLF_i$ (Series 1)	0.66	0.86	0.83	0.71	0.84	0.63
$ECLF_i$ (Series 2)	0.67	0.80	0.70	0.68	0.88	0.62
Note	When the air is supplied at floor level, series 1 is used; When the air is supplied at mid-height (about 2m above floor level), series 2 is used.					

Since the air flow in a STRAD system is complex, the stratification height sometimes may be lower than the upper boundary of the occupied zone. In these conditions, reversed flow is generated from the upper zone to the occupied zone as shown in Figure 8.1. The influence of the reversed flow to the occupied zone cooling load depends on the reversed air temperature. The higher the reversed air temperature, the more the heat contributes to the occupied zone cooling load, which frequently happens. Whereas the lower the reversed air temperature, the more the heat that will be excluded from the occupied zone cooling load. However, the reversed flow varies as the case may be and is hard to be predicted. It requires rigorous analysis of building thermal and airflow processes. Thus, an optimization process is recommended to avoid reversed flow with high air temperature by using the visualization CFD technique.

- Supply air flow rate determination

Once the occupied zone cooling load is available and the room set-point temperature is fixed, the require supply air flow rate can be determined as:

$$V_s = \frac{Q_{occupied}}{(t_{set} - t_s)\rho C_p} \quad (8\text{-}1)$$

Figure 8. 1 The STRAD system with reversed flow from upper zone to occupied zone

The room set-point temperature can be considered as the mass weight air temperature in the occupied zone or simplified as the air temperature at head level. What notable is that special attentions should be paid to the control of too large temperature difference between the heat and ankle levels, which is one of the main reasons that causes thermal discomforts in STRAD systems.

• Cooling coil load calculation

The cooling coil load for a STRAD system can be calculated according to the novel CFD-based method developed in this book. By splitting the locations of return and exhaust grilles, further energy saving potential can be achieved for the cooling coil. The energy saving for cooling coil is directly associated with the exhaust air temperature. The more the heat discharged from the exhaust grilles, the more the energy saved for the cooling coil.

• Diffusers locations

Without sacrificing the thermal comfort, the return grilles are recommended to locate at the region as low as possible, but should not be too close to the supply diffusers, which may lead to the "short-circuit" phenomenon of cold supply air. While the exhaust grilles should be located at ceiling level and close to the lighting fixture and exterior wall, which is beneficial to discharging more heat outside directly and further reducing the cooling coil load. More distributed supply diffusers are required in the space with large floor area and high ceiling level to achieve satisfied thermal environment in the occupied zone.

8. 2 Limitations and future work

This is the first time to have such an experimental study conducted in a full-scale chamber ventilated by a STRAD system with separate locations of return and exhaust grilles. However, the inside surfaces of this chamber are installed with polished aluminum plates of high reflectivity, which are different from actual building envelopes and may affect the thermal distribution in the chamber. Therefore, experimental researches in a chamber with common enclosure surfaces are called for. In addition, the simulation results in this book revealed that the height of return grille strongly affected the performance of a STRAD system, especially for the energy efficiency. However, the height of return grille in the present experimental study can only be adjusted in a very small range, which may limit the validity of our data. Thus, more flexible settings for the height of return grille are required in future studies.

Thermal stratification is formed in a space conditioned by a STRAD system. Thus, with regard to the overall efforts combined with asymmetrical radiation and local airflow, thermal comfort evaluation in such condition is more complicated than that in a uniform environment. It is recognized that the UCB thermal comfort model is applicable to evaluate the asymmetrical thermal environments. But the coupling procedure between the model and CFD simulation is manually taken and complex. An innovative coupling procedure was proposed in this book, which should be easy to operate and effective for improving evaluation accuracy. To implement this procedure, program is expected to be developed, which can automatically couple the CFD simulation results with the UCB thermal comfort model by following this simplified procedure.

In practice, a portion of exhaust air can travel through the luminaire side slots, which would discharge much convective heat generated from the lamps to the outside. However, in our simulation studies, the lamps were simplified as panels with no openings. Thus, the energy saving potentials by splitting the locations of return and exhaust grilles in a STRAD system may have been underestimated in the simulation conditions. Therefore, in future numerical studies, the effects of the exhaust function of the luminaire side slots should be considered. Furthermore, in this book, the performance of STRAD systems was only investigated with regard to thermal comfort and energy saving. The indoor air quality impact was not investigated, which directly relates to the health and productivity of the occupants. The ventilation effectiveness of STRAD systems may be affected by splitting the locations of return and exhaust grilles, which should be investigated in future studies.

References

Alajmi, A. and EI-Amer, W. Saving energy by using under-floor-air-distribution (UFAD) system in commercial buildings. *Energy Conversion and Management*. Vol. 51, pp. 637-1642 (2010)

Arens, E. A., Bauman, F. S., Johnston, L. P., Zhang, H. Testing of localized ventilation systems in a new controlled environment chamber. *Indoor Air*, Vol. 3, pp. 263-81 (1991)

Arens, E., Zhang, H., Huizenga, C. Partial- and whole-body thermal sensation and comfort, Part I: Uniform environmental conditions. *Journal of Thermal Biology*, Vol. 31, pp. 53-59 (2006)

ASHRAE Standard 62. 1-2010 Ventilation for Acceptable Indoor Air Quality. Atlanta: American Society of Heating Refrigerating and Air-Conditioning Engineers Inc., Atlanta, GA (2010)

ASHRAE ASHRAE Handbook—HVAC Applications. Atlanta: American Society of Heating Refrigerating and Air-Conditioning Engineers Inc., Atlanta, GA (2011)

ASHRAE ASHRAE Handbook—Fundamentals. Atlanta: American Society of Heating Refrigerating and Air-Conditioning Engineers Inc., Atlanta, GA (2009)

ASHRAE Standard 129-1997 (RA 2002), Measuring Air-Change Effectiveness. Atlanta: American Society of Heating Refrigerating and Air-Conditioning Engineers Inc., Atlanta, GA (1997)

Ashrae standard 55-2010 Thermal Environmental Conditions for Human Occupancy. Atlanta: American Society of Heating Refrigerating and Air-Conditioning Engineers Inc., Atlanta, GA (2010)

Awbi, H. B. Energy efficient room air distribution. *Renewable Energy*, Vol. 15, pp. 293-299 (1998)

Awad, A. S., Calay, R. K., Bardran, O. O., Holdo, A. E. An experimental study of stratified flow in enclosures. *Applied Thermal Engineering*, Vol. 28, pp. 2150-2158 (2008)

Axelrad, R. Economic implications of indoor air quality and its regulation and control. *In: Maroni, M. & Berry, M. A., ed. Pilot study on indoor air quality. The implications of indoor air quality for modern society. Report on a meeting held in Erice, Italy, February* 13-17, 1989. Washington DC, North Atlantic Treaty Organization, 1989 (CCMS Report No 183).

Azer, N. Z., Hsu S. The prediction of thermal sensation from a simple thermoregulatory model. *ASHRAE Transactions*, Vol. 83, Part 1 (1977)

Bagheri H. M. and Gorton R. L. Verification of stratified air condition design. *ASHRAE Transactions*, Vol. 93, pp. 211-227 (1987)

Balaras, C. A., Dascalaki, E., Gaglia, A., Droutsa, K, Energy conservation potential, HVAC installations and operational issues in Hellenic airports. Energy and Buildings, Vol. 35, pp. 1105-1120 (2003)

Bauman, F. S., Johnston, L. P., Zhang, H., Arens, E. A. Performance testing of a floor-based, occupant-controlled office ventilation system. *ASHRAE Transactions*, Vol. 97, pp. 553-565 (1991)

Bauman, F. S., Arens, E. A., Tanabe, S., Zhang, H. and Baharlo, A. Testing and optimizing the performance of a floor-based task conditioning system. Energy and Buildings, Vol. 22, pp. 173-86 (1995)

Bauman, F. and Webster, T. Outlook for underfloor air distribution. *ASHRAE Journal*, Vol. 43, No. 6, pp. 18-27 (2001)

Bauman, F. Underfloor air distribution (UFAD) design guide. Atlanta, GA: ASHRAE, 2003.

References

Bauman, F. , Webster, T. , Benedek, C. Cooling Airflow Design Calculations for UFAD. *ASHRAE Journal*, Vol. 49, pp. 36-44 (2007)

Bjorn, E. and Nielsen, P. V. Merging thermal plumes in the internal environment. *Proc. Healthy Buildings*, Vol. 95 (1995)

Bloomfield, L. J. and Kerr, R. C. A theoretical model of a turbulent fountain. *Journal of Fluid Mechanics*, Vol. 424, pp. 197-216 (2000)

Bonnefoy, X. , Annesi-Maesano, I. , Aznar, L. M. et al. Review of evidence on housing and health. Background document to the Fourth Ministerial Conference on Environment and Health, Budapest, Hungary, June 23-25, 2004

Brohus, H. and Nielsen, P. V. Personal exposure in displacement ventilated rooms. *Indoor Air*, Vol. 6, pp. 157-167 (1996)

Cermak, R. and Melikov, A. K. Air quality and thermal comfort in an office with under-floor, mixing and displacement ventilation. *International Journal of Ventilation*, Vol. 5, pp. 323-332 (2006)

Cermak, R. , Melikov, A. K. , Forejt, L. , Kovar, O. Performance of Personalized Ventilation in Conjunction with Mixing and Displacement Ventilation. *HVAC&R Research*, Vol. 12, pp. 295-311 (2006)

Cermak, R. , Melikov, A. K. Protection of Occupants from Exhaled Infectious Agents and Floor Material Emissions in Rooms with Personalized and Underfloor Ventilation. *HVAC&R RESEARCH*, Vol. 13, pp. 23-28 (2007)

Chen, Q. and Glicksman, L. Performance Evaluation and Development of Design Guidelines for Displacement Ventilation: Final Report to TC 5. 3 - Room Air Distribution; TC 4. 10 - Indoor Environment Modeling on ASHRAE Research Project RP-949 (1999)

Chen Q. Y. and Glicksman L. System Performance Evaluation and Design Guidelines for Displacement ventilation, Atlanta, GA: ASHRAE, 2003

Chen, Q. and Srebric, J. A procedure for verification, validation, and reporting of indoor environment CFD analyses. *HVAC&R Research*, Vol. 8, Part. 2, pp. 201-216 (2002)

Chen, Q. Y. Ventilation performance prediction for buildings: A method overview and recent applications. *Building and Environment*, Vol. 44, pp. 848-858 (2009)

Chen, Q. , Lee, K. , Mazumdar, S. , Poussou, S. , Wang, L. , Wang, M. and Zhang, Z. Ventilation performance prediction for buildings: model assessment: *Building and Environment*, Vol. 45, pp. 295-303 (2010)

Cheong, K. W. D. , Yu, W. J. , Tham, K. W. , Sekhar, S. C. and Kosonen, R. A study of perceived air quality and sick building syndrome in a field environment chamber served by displacement ventilation system in the tropics. Building and Environment, Vol. 41, pp. 1530-1539 (2006)

Cho, S. , P, Im. And Haberl, J. S. Literature review of displacement ventilation. Technical report of Energy Systems Laboratory Texas A&M University System (2005)

Cooper, P. and Linden, P. F. Natural ventilation of an enclosure containing two buoyancy sources. *Journal of Fluid Mechanics*, Vol. 311, pp. 153-76 (1996)

de Dear, R. J. , Ring, J. W. , Fanger, P. O. Thermal sensation resulting from sudden ambient temperature changes. Indoor Air, Vol. 3, pp. 181-192 (1993)

Fanger, P. O. Human requirement in future air-conditioned environments: A search for excellence. Proceedings of ISHVAC'99, Shenzhen, China, pp. 86-92 (1999)

Fiala D. Dynamic simulation of human heat transfer and thermal comfort. Ph. D. dissertation, De Montfort University, England, 1998.

Fiala, D. , Kevin, J. L. , Stohrer, M. A computer model of human thermoregulation for a wide range of environmental conditions: the passive system. Journal of Applied Physiology, Vol. 87, pp. 1957-1972 (1999)

Fiala, D., Kevin, J. L., Stohrer, M. Computer prediction of human thermoregulatory and temperature responses to a wide range of environmental conditions. International Journal of Biometeorology, Vol. 45, pp. 143-159 (2001)

Fiala, D. First principles modeling of thermal sensation responses in steady-state and transient conditions. *ASHRAE Transaction*, Vol. 109, pp. 179-186 (2003)

Fiala, D., Psikuta, A., Jendritzky, G. Physiological modeling for technical, clinical and research applications. *Frontiers in Bioscience S2*, pp. 939-968 (2010)

Filler, M. Best practices for underfloor air systems. *ASHRAE Journal*, Vol. 46, pp. 39-46 (2004)

Fisk, W. J. Health and productivity gains from better indoor environments and their relationship with building energy efficiency. *Annual Reviews of Energy & Environment*, Vol. 25, pp. 537-566 (2000)

Fong, M. L., Lin, Z., Fong, K. F., Chow, T. T., Yao, T. Evaluation of thermal comfort conditions in a classroom with three ventilation methods. *Indoor Air*, Vol. 21, pp. 231-239 (2011)

Fu, G. A transient, 3-D mathematical thermal model for the clothed human. Ph. D. Dissertation, Kansas State University, Manhattan, Kansas (1995)

An effective temperature scale based on a simple model of human physiological regulatory response. *ASHRAE Transactions*, Vol. 77, Part. 1, pp. 247-262 (1971)

Gan, G. Numerical Investigation of local discomfort in offices with displacement ventilation. *Energy and Buildings*, Vol. 23, pp. 82-157 (1995)

Gao, N. P., Niu, J. L. and Zhang, H. Coupling CFD and human body thermoregulation model for the assessment of personalized ventilation. HVAC&R Research, Vol. 12, no. 3, pp. 497-518 (2006)

Gao, N. P., Zhang, H. and Niu, J. L. Investigating Indoor Air Quality and Thermal Comfort Using a Numerical Thermal Manikin. *Indoor and Built Environment*, Vol. 16, pp. 7-17 (2007)

Givoni, B. and Goldman, R. Predicting metabolic energy cost. *Journal of Applied Physiology*, Vol. 30, pp. 429-433 (1971)

Gladstone, C. and Woods, A. W. On buoyancy-driven natural ventilation of a room with a heated floor. *Journal of Fluid Mechanics*, Vol. 441, pp. 293-314 (2001)

Gordon, A. Naked airport: A cultural history of the world's most revolutionary structure. The University of Chicago Press, Chicago, IL and London (2008)

Gordon, R. L., Bagheri, H. M. Verification of stratified air condition design, ASHRAE report 388-RP (1986).

Guan, Y. Z., Hosni, M. H., Jones, B. W., Gielda, T. P. Investigation of human thermal comfort under highly transient conditions for automobile applications - part1: experimental design and human subject testing implementation *ASHRAE Transactions*, Vol. 109, Part. 2, pp. 885-897 (2003)

Guan, Y. Z., Hosni, M. H., Jones, B. W., Gielda, T. P. Investigation of human thermal comfort under highly transient conditions for automobile applications - Part2: thermal sensation modeling. *ASHRAE Transactions*, Vol. 109, Part. 2, pp. 898-907 (2003)

Guan, Y. Z., Hosni, M. H., Jones, B. W. and Gielda, T. P. Literature review of the advances in thermal comfort modeling. ASHRAE transactions, Vol. 109, Part. 2, pp. 908-916 (2003)

Hagino, M. and Junichiro, H. Development of a method for predicting comfortable airflow in the passenger compartment. SAE Technical Paper Series, No. 922131, pp. 1-10 (1992)

Hagstroma, K., Zhivovb, A. M., Sirena, K. and Christiansonb, L. L. Influence of the floor-based obstructions on contaminant removal efficiency and effectiveness, *Building and Environment*, Vol. 37, Part. 1, pp. 55-66 (2002)

Halvonová, B. and Melikov, A. K. Performance of "ductless" personalized ventilation in conjunction with displacement ventilation: Impact of disturbances due to walking person (s). *Building and Environment*, Vol. 45, pp. 427-436 (2010)

Halvonová, B. and Melikov, A. K. Performance of "ductless" personalized ventilation in conjunction with displacement ventilation: Impact of workstations layout and partitions. *HVAC&R Research*, Vol. 16, pp. 75-94 (2010)

Halvonová, B. and Melikov, A. K. Performance of "ductless" personalized ventilation in conjunction with displacement ventilation: Impact of intake height. Building and Environment, Vol. 45, pp. 996-1005 (2010)

Han, W. P. and Gu, X. L. Application of binnacles to terminal 3 of Beijing Capital-International Airport and air distribution simulation. Journal of HV&AC, Vol. 38, pp. 115-119 (2008) (In Chinese)

Hashimoto, Y. and Yoneda, H. Numerical study on the influence of a ceiling height for displacement ventilation. The 11th International IBPSA Conference, Glasgow, Scotland, pp. 1045-1052 (2009)

He, Q. B., Niu, J. L., Gao, N. P., Z, T. and J, Z. W CFD study of exhaled droplet transmission between occupants under different strategies in a typical office room. *Building and environment*, Vol. 46, pp. 397-408 (2011)

Heinemeier, K. E., Schiller, G. E. and Benton, C. C. Task conditioning for the workplace: issues and challenges. *ASHRAE Transactions*, Vol. 96, Part 2, pp. 678-689 (1990)

Hong Kong Energy End-use data 2011. The Energy Efficiency Office Electrical & Mechanical Services Department, 2011

Hu, S. P., Chen, Q. Y. and Glicksman, L. Comparison of energy consumption between displacement and mixing ventilation systems for different U. S. buildings and climates. *ASHRAE Transactions*, Vol. 105, Part 2, pp. 453-464 (1999)

Huang, S. G. Experimental and numerical study on personalized ventilation coupled with displacement ventilation. Thesis, National University of Singapore, 2011

Huizenga, C., Zhang H, Arens E. A model of human physiology and comfort for assessing complex thermal environments. *Building and Environment*, Vol. 36, pp. 691-699 (2001)

Huizenga, C., Abbaszadeh, S., Zagreus, L. and Arens, E. Air Quality and Thermal Comfort in office Buildings: Results of a Large Indoor Environmental Quality Survey. *Proceedings of Healthy Buildings*, Lisbon, Vol. III, pp. 393-397 (2006)

Hunt, G. R., Cooper, P. and Linden, P. F. Thermal Stratification Produced by Plumes and Jets in Enclosed Spaces. *Building and Environment*, Vol. 36, No. 7, pp. 871-882 (2001)

Holmberg, S. and Chen, Q. Air flow and particle control with different systems in a classroom. Indoor air, Vol. 13, pp. 200-204 (2003)

Holmberg, R. B., Eliasson, L., Folkesson, K., Strindehag, O. Inhalation zone air quality provided by displacement ventilation. In: Proceedings of Roomvent 1990, Oslo, Session B2, pp. 32 (1990)

Howe, C. M., Holland, D. J., Livchak, A. V. Displacement ventilation — smart way to deal with increased heat gains in the telecommunication equipment room. *ASHRAE Transactions*, Vol. 109, Part. 1, pp. 323-327 (2003)

Int-Hout, D. Pressurized plenum access floor — design manual. Carrier, November (2001)

ISO standard 7730 Moderate thermal environments-determination of PMV and PPD indices and specification of the conditions for thermal comfort. Geneva: International Standards Organization (1994)

ISO/TS 14505-1, Ergonomics of the Thermal Environment-Evaluation of Thermal Environments in Vehicles-Part I: Principles and Methods for Assessment of Thermal Stress, International Standardisation Organisation, Geneva (2007)

ISO/TS 14505-2, Ergonomics of the Thermal Environment-Evaluation of Thermal Environments in Vehicles-Part II: Determination of equivalent temperature, International Standardisation Organisation, Geneva (2006).

Jones, B. W., Ogawa, Y. Transient interaction between the human body and the thermal environment. *ASHRAE Transactions*, Vol. 98, pp. 189-195 (1992)

Jones, B. W. Capabilities and limitations of thermal models for use in thermal comfort standards. Energy and Buildings, Vol. 34, pp. 653-659 (2002)

Karimipanah, T., Sandberg, M. and Awbi, H. B. A comparative study of different air distribution systems in a classroom. *In: Proceedings of Roomvent* 2000, Vol. 2, Reading, UK, pp. 1013-1018 (2000)

Kaye, N. B. and Hunt, G. R. The effect of floor heat source area on the induced airflow in a room. *Building and Environment*, Vol. 45, pp. 839-847 (2010)

Kegel, B. and Schulz, U. W. Displacement ventilation for office buildings. Proceedings of 10th AIVC Conference, Vol. 1 (1989)

Kim, K. H, Kang, S. K., Lee, J. H., Oh, M. D. Analysis of thermal environment in an airport passenger terminal.

Numerical Heat Transfer, Part A: Applications: An International Journal of Computation and Methodology, Vol. 40, pp. 531-541 (2001)

Kobayashi, N. and Q, C. Y. Floor supply displacement ventilation in a small office. *Indoor and Built Environment*, Vol. 12, pp. 281-291 (2003)

Kohri, I. and Moschida, T. Evaluation method of thermal comfort in a vehicle with a dispersed two-node model. Part 1—development of dispersed two-node model. *Journal of Human-Environmental System*, Vol. 6, Part. 1, pp. 19-29 (2002)

Kohri, I., Moschida, T. Evaluation method of thermal comfort in a vehicle with a dispersed two-node model. Part 2—development of new evaluation. *Journal of Human-Environmental System*, Vol. 6, Part. 2, pp. 77-91 (2003)

Kuo, J. Y. and Chung, K. C. The effect of diffuser's location on thermal comfort analysis with different air distribution strategies. Journal of Building Physics, Vol. 22, pp. 208-229 (1999)

Lam, J. C. and Chan, A. L. S. CFD analysis and energy simulation of a gymnasium. Building and Environment, Vol. 36, pp. 351-358 (2001)

Lau, J. and Niu, J. L. Measurements and CFD simulation of the temperature stratification in an atrium using a floor level air supply method. *Indoor and Built Environment*, Vol. 12, no. 4, pp. 265-80 (2003)

Lee, K. S., Zhang, T., Jiang, Z. and Chen, Q. Y. Air distribution effectiveness with stratified air distribution systems. *ASHRAE Transactions*, Vol. 115, no. 2 (2009)

Lee, K. S., Zhang, T., Jiang, Z., and Chen, Q. Comparison of airflow and contaminant distributions in rooms with traditional displacement ventilation and under-floor air distribution systems, ASHRAE Transactions, Vol. 115, Part 2. (2009)

Lehrer, D. and Bauman, F. Hype vs. Reality: New Research Findings on Underfloor Air Distribution Systems. Proceedings of Greenbuild 2003, Pittsburgh, PA (2003)

Li, X. T. , Wang, X. , Li, X. F. , Li, Y. Investigation on the relationship between flow pattern and air age. The 6th International IBPSA Conference—Building Simulation' 99, Kyoto, Japan, pp. 423-429 (1999)

Li, Q. , Yoshino, H. , Mochida, A. , Lei, B. , Meng, Q. L. , Zhao, L. H. , Lun, Y. CFD study of the thermal environment in an air-conditioned train station building. Building and Environment, Vol. 44, pp. 1452-1456 (2009)

Li, R. X. , S, S. C. and M, A. K. Thermal comfort and IAQ assessment of underfloor air distribution system integrated with personalized ventilation in hot and humid climate. *Building and Environment*, Vol. 45, pp. 1906-1913 (2010)

Lian, Z. and Wang, H. Experimental study factors that affect thermal comfort in an upward-displacement air conditioned room. *HVAC & R Research*, Vol. 8, Part. 2, pp. 191-200 (2002)

Lin, Z. , Chow, T. T. , Fong, K. F. , Tsang, C. F. , Wang, Q. W. Comparison of performances of displacement and mixing ventilations. Part II: Indoor air quality, *International Journal of Refrigeration*, Vol. 28, pp. 288-305 (2005)

Lin, Z. , Chow, T. T. , Fong, K. F. , Tsang, C. F and Wang, Q. Comparison of performances of displacement and mixing ventilations. Part I: Thermal comfort. *International Journal of Refrigeration*, Vol. 28, pp. 288-305 (2005)

Lin, Z. , Chow, T. T. , Tsang, C. F. , Fong, K. F. , Chan, L. S. CFD study on effect of the air supply location on the performance of the displacement ventilation system. Building and Environment, Vol. 40, pp. 1051-1067 (2005)

Lin, Z. , Lee, C. K. , Fong, S. , Chow, T. T. , Yao, T. , Chan, A. L. S. Comparison of annual energy performances with different ventilation methods for cooling. Energy and Buildings, Vol. 43 (1), pp. 130-136 (2011)

Lin, Z, Lee, C. K. , Fong, K. F. , Chow, T. T. Comparison of annual energy performances with different ventilation methods for temperature and humidity control. Energy and Buildings, Vol. 43 (1), pp. 3599-3608 (2011)

Lin, Zhang, Chow, T. T. , Tsang, C. F. , Fong, K. F. , Chan, L. S. Effects of headroom on the performance of the displacement ventilation system. *Indoor and Built Environment*, Vol. 15, pp. 333-346 (2006)

Lin, Y. J. P. and Linden, P. F. A model for an under floor air distribution system. *Energy and Buildings*, Vol. 37, pp. 399-409 (2005)

Linden, P. F. , Lane-Serff, G. F. , Smeed, D. A. Emptying filling boxes, the fluid mechanics of natural ventilation. *Journal of Fluid Mechanics*, Vol. 212, pp. 309-35 (1990)

Liu, Q. A. The fluid mechanics of underfloor air distribution. Ph. D. thesis, University of California, San Diego (2005)

Liu Y. and Hiraoka, K Ventilation Rate requirement for Raised Floor HVAC System. *SHASE Transactions*, Vol. 67, pp. 75-84 (1997)

Liu, Q. A. and Linden, P. L. The fluid mechanics of underfloor air distribution. *Journal of Fluid Mechanics*, Vol. 554, pp. 323-341 (2006)

Lofeness, V. , Brahme, R. , Mondazzi, M. , Vineyard, E. and MacDonald, M. Technical Report, Center for Building Performance and Diagnostics, Pittsburgh, PA (US); Oak Ridge National Laboratory, Oak Ridge, TN (US) (2002)

Loudermilk, K. J. Underfloor air distribution solutions for open office applications. *ASHRAE Transactions*, Vol. 105, pp. 605-13

Makhoul, A., Ghali, K. and Ghaddar, N. A simplified combined displacement and personalized ventilation model. *HVAC&R Research*, Vol. 18, pp. 737-749 (2012)

Makhoul, A., Ghali, K. and Ghaddar, N. The energy saving potential and the associated thermal comfort of displacement ventilation systems assisted by personalised ventilation. *Indoor and Built Environment*, Published online before print, doi: 10.1177/1420326X12443847 (2012)

Matsunaga, K., Sudo, F., Tanabe, S, Madsen, T. L. Evaluation and measurement of thermal comfort in the vehicles with a new thermal manikin. SAE Paper Series, No. 931958, pp. 35-43 (1993)

Mattson, M. and Sandberg, M. Displacement ventilation — Influence of physical activity. In: Proceedings of Roomvent, Fourth International Conference on Air Distribution in Rooms, Crakow, Poland, Silesian Technical University, Vol. 2, pp. 77-92 (1994)

McDonell, G. Underfloor & Displacement: Why they're not the same. *ASHRAE Journal*, Vol. 45, pp. 18-22

McQuiston, F. C. and Parker, J. D. Spitler Heating, Ventilating and Air Conditioning Analysis and design. John Wiley & Sons, Inc., New York, 1994

Melikov A. K. and Nielsen J. B. Local thermal discomfort due to draft and vertical temperature difference in rooms with displacement ventilation. *ASHRAE Transactions*, Vol. 96, pp. 1050-1057 (1989)

Melikov, A and Zhou, H. Q. Comparison of methods for determining teq under well-defined conditions. *In: Proceedings of the Sixth International Conference Florence ATA*, Florence, Italy (1999)

Mundt, E. Convection flows in rooms with temperature gradients — theory and measurements. Roomvent' 92, Third International Conference, Aalborg, Denmark, Publisher: DANVAK, Lyngby, Denmark, Vol. 3, pp. 69-86 (1992)

Mundt, E. Displacement ventilation systems-convection flows and temperature gradients. *Building and Environment*, Vol. 30, pp. 129-133 (1995)

Mundt, E. The performance of displacement ventilation system. Ph. D. thesis, Royal institute of Technology, Sweden (1996)

Nielsen, P. V. Displacement Ventilation — Theory and Design. Aalborg University, Aalborg (1993)

Nielsen, P. V. Temperature distribution in a displacement ventilated room. *Proceedings of Roomvent'*96, 5th *International Conference on Air Distribution in Rooms*, July 17-19, Yokohama, Japan, Vol. 3, pp. 323-330 (1996)

Nilsson, H. O. and Holmér I. Comfort climate evaluation with thermal manikin methods and computer simulation models. *Indoor Air*, Vol. 13, pp. 28-37 (2003)

Nilsson, H. O. Thermal comfort evaluation with virtual manikin methods. Building and Environment, Vol. 42, pp. 4000-4005 (2007)

Nielsen, P. V., Larsen, T. S. and Topp, C. Design methods for air distribution systems and comparison between mixing ventilation and displacement ventilation. *In: Proceedings of Healthy Buildings*, Singapore, Vol. 2, pp. 492-497 (2003)

Nielsen, P. V., Allard, F., Awbi, H. B., Davidson, L. and Schälin, A. Computational fluid dynamics in ventilation design. REHVA Guidebook No. 10, printed in Finland, Forssan Kirjapaino Oy, Forssa (2007)

Novoselac, A. and Srebric, J. A cirtical review on the performance and design of combined cooled ceiling and displacement ventilation systems. *Energy and Buildings*, Vol. 34, pp. 497-509 (2002)

Novoselac, A. and Srebric, J. Comparison of air exchange efficiency and contaminant removal effectiveness as IAQ indices. *ASHRAE Transactions*, Vol. 109, Part. 2, pp. 339-349 (2003)

Olesen, B. W., Koganei, M., Holbrook, G. T. and Woods, J. E. Evaluation of a vertical displacement ventilation system. *Building and Environment*, Vol. 29, pp. 303-310 (1994)

Olivieri, J. B. and Singh, T. Effect of supply and return air outlets on stratification/energy consumption. *ASHRAE Transaction*, Vol. 188, Part. 1 (1982)

Code of Practice for Overall Thermal Transfer Value in Buildings, Building Department, Hong Kong, April 1995.

Park, H. j. and Holland, D. The effect of location of a convective heat source on displacement ventilation: CFD study, *Building and Environment*, Vol. 36, pp. 883-889 (2001)

Parker, J., Cropper, P. and Li, S. Using building simulation to evaluation low carbon refurbishment options for airport buildings. *Proceedings of building simulation* 2011, 12th conference of international building performances simulation association, Sydney (2011)

Posner, J. D., Buchanan, C. R. and Dunn-Rankin, D. Measurement and prediction of indoor air flow in a model room. *Energy and Buildings*. Vol. 35, pp. 515-526 (2003)

Skistad, H. (ed), Mundt, E., Nielsen, P. V., Hagström, K. AND Railio, J. Displacement ventilation in non-industrial premises. REHVA Guidebook, No. 1 (2002)

Ring, J. W. and de Dear, R. J. Temperature transients: a model for heat diffusion through the skin, thermoreceptor response and thermal sensation. Indoor air, Vol. 1, pp. 448-456 (1990)

Sandberg, M. and Blomqvist, C. Displacement ventilation systems in office rooms. *ASHRAE Transactions*, Vol. 95, Part. 2 (1989)

Schiavon, S., Melikov, A., Cermak, C., De Carli, M. and Li X. An Index for Evaluation of Air Quality Improvement in Rooms with Personalized Ventilation Based on Occupied Density and Normalized Concentration. *Proceedings of International Conference Roomvent* 2007, Helsinki, Finland, C05, PP. 1294 (2007)

Schiavon, S., Lee, K. H., Bauman, F., Webster, T. Simplified calculation method for design cooling loads in underfloor air distribution (UFAD) systems. *Energy and Buildings*, Vol. 43, pp. 517-528 (2010)

Seppanen, O. A., Fisk, W. J., Eto, J., Grimsrud, D. T. Comparison of conventional mixing and displacement air-conditioning and ventilation systems in U. S. commercial buildings. *ASHRAE Transactions*, Vol. 95, No. 2, pp. 1028-1040 (1989)

Sekhar, S. C., Gonga, N., Tham, K. W., Cheong, K. W., Melikov, A. K., Wyon, D. P., Fanger, P. O. Findings of personalized ventilation studies in a hot and humid climate. *HVAC&R Research*, Vol. 11, pp. 603-620 (2005)

Sekhar, S. C. and Ching, C. S. Indoor air quality and thermal comfort studies of an under-floor air-conditioning system in the tropics. *Energy and Buildings*, Vol. 34, pp. 431-44 (2001)

Shaughnessy, R. J., Haverinen-Shaughnessy, U., Nevalainen, A., Moschandreas, D. A preliminary study on the association between ventilation rates in classrooms and student performance. *Indoor Air*, Vol. 16, pp. 465-468 (2006)

Simmonds, P. Creating a micro-climate in a large airport building to reduce energy consumption. ASHRAE Conference on Buildings in Hot and Humid Climates. Ft. Worth (1996)

Simmonds, p., Holst, S., Reuss, S., Gaw, W. Comfort conditioning for large spaces. P. Simmonds. *ASHRAE Transactions*. Vol. 105, Part. 1, pp. 1037-1048 (1999)

Smith, C. E. A transient, three-dimensional model of human thermal system. PH. D. Dissertation, Kansas State University, Manhattan, Kansas (1991)

Skistad, H. Displacement Ventilation, Taunton, UK: Research Studies Press Ltd, 1994.

Skistad, H. , Mundt, E. , Nielsen, P. V. , Hagström, K. and Railio J. Displacement Ventilation in Non-industrial Premises, REHVA, Brussels: Guidebook no. 1 (2002)

Sodec, F. and Craig, R. Under-floor air supply system the European experience. *ASHRAE Transactions*. Vol. 96, Part. 2, pp. 690-695 (1990)

Stymne, H. , Sandberg, M. and Mattson, M. Dispersion pattern of contaminants in a displacement ventilation room. Proceedings of the 12th AIVC Conference, Ottawa, Canada, Vol. 1, pp. 173-190 (1991)

Stolwijk, J. A. J. A mathematical model of physiological temperature regulation in man. NASA contractor report, NASA CR 1855, Washington, DC (1971)

Sun, W. M. , Cheong, K. W. D. , Melikov, A. k. Subjective study of thermal acceptability of novel enhanced displacement ventilation system and implication of occupants' personal control. *Building and Environment*, Vol. 57, pp. 49-57 (2012)

Svensson, A. G. L. Nordic experiences of displacement ventilation systems. *ASHRAE Transactions*, Vol. 95, Part. 2, pp. 1013-1017 (1989)

Tanabe, S. , Kobayashi, S. , Nakano, J. , Ozeke, Y. , Konishi, M. Evaluation of thermal comfort using combined multi-node thermoregulation (65MN) and radiation models and computational fluid dynamics (CFD) . Energy and Buildings Vol. 34, pp. 637-646 (2002)

Taniguchi, Y. , Hiroshi, A. and Kenji, F. Study on car air conditioning system controlled by car occupants' skin temperatures— Part 1: Research on a method of quantitative evaluation of car occupants' thermal sensations by skin temperature. SAE Technical Paper Series, No. 920169, pp. 13-19 (1992)

Topp, C. , Nielsen, P. V. , Sorensen, D. N. , Application of computer simulated persons in indoor environmental modeling. *ASHRAE Transactions*, Vol. 108, pp. 1084-1089 (2002)

Tian, L. , Lin, Z. and Wang, Q. W. Comparison of gaseous contaminant diffusion under stratum ventilation and under displacement ventilation. *Building and Environment*, Vol. 45, pp. 2035-2046 (2010)

Thimijan, R. W. and Heins, R. D. Photometric, radiometric, and quantum light units of measure: a review of procedures for interconversion. *HortScience*, Vol. 18, pp. 818-822 (1983)

Walgama, C. , Fackrell, S. , Karimi, M. , Fartaj, A. and Rankin, G. W. Passenger thermal comfort in vehicles — a review. Proc. IMechE. Vol. 220, Part D: J. Automobile Engineering, pp. 543-562 (2006)

Wang, X. L. Thermal comfort and sensation under transient conditions. Ph. D. dissertation, Department of Energy Technology, Division of Heating and Ventilation, The Royal Institute of Technology, Sweden, 1994.

Wan, M. P. and Chao, Christopher Y. Experimental study of thermal comfort in an office environment with an underfloor ventilation system. *Indoor and Built Environment*, Vol. 11, no. 5, pp. 250-265 (2002)

Wan, M. and Chao, C. Numerical and Experimental Study of Velocity and Temperature Characteristics in a Ventilated Enclosure with Underfloor Ventilation Systems. *Indoor Air*, Vol. 15, No. 5, pp. 342-355 (2005)

Wargocki, P. , Wyon, D. P. , Matysiak, B. , Irgens, S. The effects of classroom air temperature and outdoor air supply rate on the performance of school work by children. *In*: *Proceedings of the* 10th *International Conference on IAQ and Climate*, Beijing, China (2005)

Webster, T. , Bauman, F. , Shi, M. Y. and Rees, J. Underfloor air distribution: thermal stratification. *ASHRAE Journal*, Vol. 44, pp. 28-36 (2002)

Webster, T. , Bannon, R. and Lehrer, D. Teledesic Broadband Center Field Study, Center for the Built Environment, University of California, Berkeley, USA (2002)

Wissler, E. H., Mathematical simulation of human thermal behavior using whole body models. In: Heat Transfer in Medicine and Biology. A. Shitzer and R. C. Eberhart Eds, Plenum Press, New York, pp. 325-373 (1985)

Wyon, D., Larsson, S., Forsgren, B., Lundgren, I. Standard procedures for assessing vehicle climate with a thermal manikin. SAE Technical Paper Series, No 890049, pp. 1-11 (1989)

Wyon, D. P. and Sandberg M. Thermal manikin prediction of discomfort due to displacement ventilation. *ASHRAE Transactions*, Vo. 96, Part. 1, pp. 67-75 (1990)

Xing, H., Hatton, A., Awbi, H. B. A study of the air quality in the breathing zone in a room with displacement ventilation. *Building and Environment*, Vol. 36, pp. 809-820 (2001)

Xu, H. T., Gao, N. P. and Niu, J. L. A Method to Generate Effective Cooling Load Factors for Stratified Air Distribution Systems Using a Floor-Level Air Supply, *HVAC&R Research*, Vol. 15, pp. 915-930 (2009)

Xu, H. T. and Niu, J. L. Numerical procedure for predicting annual energy consumption of the under-floor air distribution system. *Energy and Buildings*, Vol. 38, pp. 641-647 (2006)

Yasuyuki Miyagawa, The heat loads of air conditioning systems in large space Building equipment and plumbing, special issue, Vol. 10, pp. 29-36 (1983) (in Japanese)

Yu, B. F., Hu, Z. B., Liu, M., Yang, H. L., Kong, Q. X. and Liu, Y. H. Review of research on air-conditioning system and indoor air quality control for human health. *International Journal of Refrigeration*, Vol. 32, pp. 3-20 (2009)

Yu, W. J., Cheong, K. W. D., Sekhar, S. C., Tham, K. W., Kosonen, R. Local discomfort caused by draft perception in a space served by displacement ventilation system in the tropics. *Indoor and Built Environment*, Vol. 15, no. 3, pp. 225-233 (2006)

Yuan, X., Chen, Q. Y., Glicksman, L. R. A critical review of displacement ventilation. *ASHRAE Transaction*, Vol. 104, pp. 78-90 (1998)

Zhang, H., Huizenga, C., Arens, E., Yu, T. F. Considering individual physiological differences in a human thermal model. *Journal of Thermal Biology*, Vol. 26, pp. 401-408 (2001)

Zhang, H. Human thermal sensation and comfort in transient and non-uniform thermal environments, Ph. D. Thesis, University of California Berkeley 2003, 415pp.

Zhang, H., Huizenga, C., Arens, E. and Yu, T. Modeling thermal comfort in stratified environments. *The 10th International Conference on Indoor Air Quality and Climate*, Beijing, China, pp. 133-137 (2005)

Zhang H, Arens E, Huizenga C. Thermal sensation and comfort models for non-uniform and transient environments: Part I: local sensation of individual body parts. *Building and Environment*, Vol. 45, pp. 380-388 (2010)

Zhang H, Arens E, Huizenga C. Thermal sensation and comfort models for non-uniform and transient environments: Part II: local comfort of individual body parts. *Building and Environment*, Vol. 45, pp. 389-398 (2010)

Zhang H, Arens E, Huizenga C, Han T. Thermal sensation and comfort models for non-uniform and transient environments: Part III: Whole-body sensation and comfort. *Building and Environment*, Vol. 45, pp. 399-410 (2010)

Zhang, Z. and Chen, Q. Experimental measurements and numerical simulations of particle transport and distribution in ventilated rooms. *Atmospheric Environment*, Vol. 40, pp. 3396-3408 (2006)

Zhang, T. and Chen, Q. Novel air distribution systems for commercial aircraft cabins. *Building and Environment*, Vol. 42, pp. 1675-1684 (2007)